T0256875

Bayesian Non- and Semi-parametric
Methods and Applications

THE ECONOMETRIC AND TINBERGEN
INSTITUTES LECTURES

Series Editors
Herman K. van Dijk and Philip Hans Franses
The Econometric Institute,
Erasmus University Rotterdam

The Econometric Institute Lectures series is a joint project of Princeton University Press and the Econometric Institute at Erasmus University Rotterdam.

This series collects the lectures of leading researchers which they have given at the Econometric Institute for an audience of academics and students.

The lectures are at a high academic level and deal with topics that have important policy implications. The series covers a wide range of topics in econometrics. It is not confined to any one area or sub-discipline.

The Econometric Institute is the leading research center in econometrics and management science in the Netherlands. The Institute was founded in 1956 by Jan Tinbergen and Henri Theil, with Theil being its first director. The Institute has received worldwide recognition with an advanced training program for various degrees in econometrics.

Other books in this series include

Anticipating Correlations: A New Paradigm for Risk Management by Robert Engle

Complete and Incomplete Econometric Models by John Geweke

Social Choice with Partial Knowledge of Treatment Response by Charles F. Manski

Yield Curve Modeling and Forecasting: The Dynamic Nelson-Siegel Approach by Francis X. Diebold and Glenn D. Rudebusch

Bayesian Non- and Semi-parametric Methods and Applications

Peter E. Rossi

Princeton University Press

Princeton and Oxford

Published by Princeton University Press, 41 William Street, Princeton, New Jersey 08540
In the United Kingdom: Princeton University Press, 6 Oxford Street, Woodstock, Oxfordshire OX20 1TW

press.princeton.edu

Library of Congress Cataloging-in-Publication Data

Rossi, Peter E. (Peter Eric), 1955–
  Bayesian non- and semi-parametric methods and applications / Peter E. Rossi.
    pages cm.—(The Econometric and Tinbergen Institutes lectures)
  Includes bibliographical references and index.
  ISBN 978-0-691-14532-7 (hardcover : alk. paper) 1. Econometrics. 2. Bayesian statistical decision theory. 3. Mathematical economics. I. Title.
  HB139.R64 2014
  330.01'519542—dc23
                          2013038609

British Library Cataloging-in-Publication Data is available

This book has been composed in Sabon Next LT Pro

Printed on acid-free paper. ∞

Typeset by S R Nova Pvt Ltd, Bangalore, India

Printed in the United States of America

10 9 8 7 6 5 4 3 2 1

Contents

Preface   vii

1   Mixtures of Normals   1
   1.1   Finite Mixture of Normals Likelihood
       Function   6
   1.2   Maximum Likelihood Estimation   9
   1.3   Bayesian Inference for the Mixture of
       Normals Model   15
   1.4   Priors and the Bayesian Model   16
   1.5   Unconstrained Gibbs Sampler   25
   1.6   Label-Switching   29
   1.7   Examples   34
   1.8   Clustering Observations   46
   1.9   Marginalized Samplers   49

**2   Dirichlet Process Prior and Density
Estimation**   59
   2.1   Dirichlet Processes—A Construction   60
   2.2   Finite and Infinite Mixture Models   64
   2.3   Stick-Breaking Representation   68
   2.4   Polya Urn Representation and Associated
       Gibbs Sampler   70
   2.5   Priors on DP Parameters and
       Hyper-parameters   72
   2.6   Gibbs Sampler for DP Models and Density
       Estimation   78
   2.7   Scaling the Data   80
   2.8   Density Estimation Examples   81

3   Non-parametric Regression   90

    3.1   Joint vs. Conditional Density Approaches   90
    3.2   Implementing the Joint Approach with
          Mixtures of Normals   93
    3.3   Examples of Non-parametric Regression
          Using Joint Approach   96
    3.4   Discrete Dependent Variables   104
    3.5   An Example of Expenditure Function
          Estimation   108

4   Semi-parametric Approaches   115

    4.1   Semi-parametric Regression with DP Priors   115
    4.2   Semi-parametric IV Models   122

5   Random Coefficient Models   152

    5.1   Introduction   152
    5.2   Semi-parametric Random Coefficient Logit
          Models   157
    5.3   An Empirical Example of a Semi-parametric
          Random Coefficient Logit Model   161

6   Conclusions and Directions
    for Future Research   187

    6.1   When Are Non-parametric and
          Semi-parametric Methods Most Useful?   187
    6.2   Semi-parametric or Non-parametric
          Methods?   189
    6.3   Extensions   191

Bibliography   195

Index   201

Preface

Much of the work in Bayesian econometrics has focused on showing the value of Bayesian methods for parametric models (see, for example, Geweke (2005), Koop (2003), Li and Tobias (2011), and Rossi, Allenby, and McCulloch (2005)). This literature has documented the superior sampling properties of Bayesian parametric methods, particularly in situations with limited sample information. For example, Bayesian hierarchical models have become standard in the marketing literature because of the prevalence of panel data structures and consumer heterogeneity. While these models that have proved so useful in micro-econometrics and marketing applications involve very large numbers of parameters, both the panel-unit level model (typically a linear regression or multinomial logit specification) and the distribution of heterogeneity (the random "effects" distribution—commonly assumed to be multivariate normal) are parametric. The purpose of this book is to show how Bayesian non-parametric methods can be applied to econometric and marketing applications. The Bayesian non-parametric methods used here are finite and infinite mixture models and the central theme is to express the non-parametric problem as fundamentally a problem of density approximation/estimation.

A Bayesian approach to non-parametric problems is fundamentally what I will call a "full-information" approach. That is, all Bayesian approaches are likelihood-based and require an approximation to the joint distribution of all observables. This contrasts to the approach often favored in the non-Bayesian econometrics literature which I will call "partial-information." In the "partial-information" approach (often the basis for GMM methods), only certain assumptions (such as independence or orthogonality restrictions) are used to form the estimator of the model parameters and no attempt is made to model the entire distribution of the data. In the "partial-information" approach, the sampling distribution of the estimator is derived (typically via asymptotic methods) without making explicit distributional assumptions. The resulting inference is often claimed to be "distribution-free" for this reason. However, it is well recognized that there can be considerable efficiency losses from this approach. The argument that is used to justify the approach is that it does not require arbitrary and unsubstantiated distributional assumptions. Bayesian parametric approaches are then criticized on the basis that they require arbitrary distribution assumptions which often are not examined. Concerns for model specification cause the "partial-information" advocate to be willing to trade-off efficiency for robustness of inference.

In principle, a fully non-parametric Bayesian approach should remove concerns regarding model mis-specification while retaining the desirable sampling properties of a "full-information" or likelihood-based procedure. As a practical matter, however, the full non-parametric approach may require the approximation of high dimensional distributions. Bayesian approaches will essentially involve very flexible or highly parameterized models that can be given a non-parametric interpretation. Non-Bayesian procedures for highly parameterized models often suffer from the over-fitting problem where small subsets of data drive the model fit and the model produces very non-smooth estimates. The advantage of a Bayesian method in a highly parameterized model is that, with proper priors, the tendency to "over-fit" is reduced substantially. This is because, in many models, proper priors impose what amounts to a penalty for highly parameterized models and implement a form of shrinkage.[1] For Bayesian procedures, the concerns regarding over-fitting are replaced with the problem of assessment of proper priors. There has been insufficient attention devoted to the assessment of priors in Bayesian non-parametric applications. The assumption often made is that quality of Bayesian density approximation is

---

[1] Here the term "shrinkage" refers to the property of Bayes estimators with proper priors to be somewhere between the location of the prior and the likelihood, thereby reducing the sensitivity of the estimators to sampling error.

not influenced a great deal by the prior settings. There is a sense in which this statement is obviously false—there certainly are prior settings which exert a great deal of influence and limit the flexibility of the Bayesian procedure. More importantly, some of the standard, proper but diffuse, settings used in practice can be highly informative in ways that, perhaps, are unintended. What is required is a procedure for prior assessment that retains flexibility and smoothing. I find that it is relatively straightforward to avoid overly and, perhaps, unintentionally informative priors and find a region of prior settings where prior sensitivity is limited and yet the procedure retains desirable smoothing properties.

In the non-Bayesian literature on non-parametric approaches, the style of research is to propose a "non-parametric" procedure and then prove that as the sample size approaches infinity this procedure "consistently" (based on some norm which measures the difference between the non-parametric approximation and the "true" model) recovers the true model. In some cases, further analysis is done to determine the optimal rate at which parameters must be added to the "non-parametric" approximation and to provide some sort of asymptotic distribution theory. Of course, any number of approaches are consistent in the sense that there are many different possible sets of basis functions which can be used to approximate an arbitrary distribution. In some cases, some sort of theory of "testing" or a penalty function is used to expand the model as the sample size increases at a rate sufficient to assure consistency but not too rapidly to avoid overfitting. For the user of the procedure, there is little guidance as to how to apply the approximation to one sample of a fixed size. Cross-validation methods are sometimes used to "tune" the approximation for a given sample, but have unknown finite sample properties.

In the approach taken here, I will use mixtures of normals as the basis for approximations. It is obvious that mixtures of normals have the desired "consistency" property required of a non-parametric approach. What is more important is to demonstrate that the Bayesian procedures provide reasonable estimates for the demanding sorts of examples considered in applied work. With proper procedures for prior assessment, the "flexibility" vs "over-fitting" problem is largely avoided. Approximations involving literally thousands of parameters can be used without concern. The Bayes Factors[2] implicit in Bayesian inference impose such strong smoothing and shrinkage properties that I do not observe the over-fitting problem. This means that I don't have to couple my inference procedure with additional (often ad hoc) methods for avoiding the over-fitting problem. This provides a real advantage to a thoughtful Bayesian approach.

There is a sense, however, that all non-parametric methods, no matter how powerful, are very demanding of the data. Ultimately, inference regarding densities demands a uniform distribution or lack of sparseness in the data. It is not so much the number of observations that matters, but more that the observations are spread evenly across the, possibly high dimensional, space. For sparse data or data with "holes," non-parametric methods will not work as well. The great advantage of Bayesian procedures is that the inference procedure produces reliable information regarding the precision of inference regarding the density and any functional thereof. This inference comes essentially free as part of the Bayes computing procedure and without other delta method or bootstrap computations. The inference will show that in sparse areas of the space, the density is imprecisely estimated. Although it is reassuring to know that the Bayesian procedure will properly reflect lack of sample information, this does not remove the problem which is particularly acute with high dimensional distributions. For this reason, interest may focus on semi-parametric Bayesian methods. In semi-parametric methods, part of the problem is modeled with parametric assumptions and non-parametric

---

[2]The Bayes Factor is the quantity by which one transforms the prior probability of a model to its posterior probability. In any parametric model, the Bayes Factor can be expressed as $\int p\,(Data|\theta)\,p\,(\theta)\,d\theta$, where $p\,(Data|\theta)$ is the distribution of the observed data given the parameters (the "model") and $p\,(\theta)$ is the prior.

methods are employed for the remaining smaller dimensional part of the problem. Linear index models are a good example of a semi-parametric approach. For example, consider the problem of modeling the joint distribution of (y, x). If this is high dimensional, the semi-parametric approach would be to model the bivariate joint distribution of $(y, x'\gamma)$ using non-parametric density approximations. In this semi-parametric approach, the linear index serves as a dimension reduction device (see Chapter 4 for further discussion).

Panel data provide a good example of a natural setting where semi-parametric approaches are desirable. Invariably, panels involve a large number of cross-sectional units tracked over a limited duration of time. Here the time dimension is insufficient to allow for non-parametric approaches to the model at the cross-sectional level, but there are sufficient cross-sectional units to allow for a non-parametric determination of a random coefficient distribution. For example, with purchase data, we might specify a standard multinomial logit model for each unit indexed by a parameter vector which has an arbitrary distribution across units (see Chapter 5).

Ultimately, part of the output of a non-parametric or semi-parametric approach will be a density estimate or some functional of this object. For example, a non-parametric random coefficient model will yield a joint distribution of preference parameters across consumers. This joint distribution will be represented by a high dimensional density surface. A challenge will be to summarize this density in a meaningful way. The ability to compute univariate and bivariate marginals from this distribution will become important and this is another advantage of the mixture approach in that the implied marginal distributions are easy to calculate. Visualization techniques will play an important role in summarizing these distributions as moments lose much of their significance or interpretability for non-elliptical distributions.

I develop all of the methods here from first principles so that this work is accessible to anyone with a reasonable introductory knowledge of Bayesian statistics. Specifically, I assume that readers are familiar with the Bayesian paradigm, Bayesian inference for multivariate regression models with normal errors, and the Gibbs sampler. This material is covered well in many texts including Rossi, Allenby, and McCulloch (2005) (which may be preferred due to notational compatibility).

In this book, all methods are illustrated with both simulated and actual data. In addition, the finite and infinite mixture approaches are implemented in my contributed R (R (2012)) package, *bayesm* (Rossi (2012)). Given the modular properties of MCMC methods, it will be a simple matter to use the *bayesm* routines to implement a computational method for the many possible models which can be crafted from a mixture of normals approach.

Summary

Most econometric models used in micro-economics and marketing applications involve arbitrary distributional assumptions. For example, the standard normal linear regression model assumes that the error terms in a regression are normally distributed and that the regression function is linear. Another important example is the use of the multivariate normal distribution as a model for heterogeneity or for the distribution of parameters across different units in a panel data structure. As more and less sparse data becomes available, there is a natural interest to provide methods which relax these distributional assumptions. In the Bayesian approach advocated here, specific distributional assumptions are replaced with more flexible distributions based on mixtures of normals. The Bayesian can use either a large but fixed number of normal components in the mixture or an infinite number bounded only by the sample size. By using flexible distributional approximations instead of fixed parametric models, the Bayesian can reap the advantages of an efficient method which models all of the structure in the data while retaining desirable smoothing properties. Non-Bayesian non-parametric methods often require additional ad hoc rules to avoid "over-fitting" in which resulting density approximates are non-smooth. With proper priors, the Bayesian approach largely avoids over-fitting, while retaining flexibility. I provide methods for assessing informative priors that only require simple data normalizations. I apply the mixture of normals approximation method to a number of important models in micro-econometrics and marketing including the general regression model, instrumental variable problems, and models of heterogeneity.

Acknowledgments

I'd like to thank Herman van Dijk for inviting me to give the PUP lectures at Erasmus University. Thanks also to Sanjog Misra, Nick Polson, and Matt Taddy for many useful discussions. My co-authors, Jean-Pierre Dube, Guenter Hitsch, Chris Hansen, Tim Conley, and Rob McCulloch, are responsible for much of this material and I am very grateful for all of the insights I have obtained over the years by working with them. I am also very grateful to Renna Jiang for excellent research assistance. I am grateful for very detailed comments from Professor Dennis Fok and an anonymous reviewer that improved the readability of the manuscript.

Bayesian Non- and Semi-parametric
Methods and Applications

1

Mixtures of Normals

In this chapter, I will review the mixture of normals model and discuss various methods for inference with special attention to Bayesian methods. The focus is entirely on the use of mixtures of normals to approximate possibly very high dimensional densities. Prior specification and prior sensitivity are important aspects of Bayesian inference and I will discuss how prior specification can be important in the mixture of normals model. Examples from univariate to high dimensional will be used to illustrate the flexibility of the mixture of normals model as well as the power of the Bayesian approach to inference for the mixture of normals model. Comparisons will be made to other density approximation methods such as kernel density smoothing which are popular in the econometrics literature.

The most general case of the mixture of normals model "mixes" or averages the normal distribution over a mixing distribution.

$$p\left(y|\tau\right) = \int \phi\left(y|\mu, \Sigma\right) \pi\left(\mu, \Sigma|\tau\right) d\mu d\Sigma \qquad (1.0.1)$$

Here $\pi(\ )$ is the mixing distribution. $\pi(\ )$ can be discrete or continuous. In the case of univariate normal mixtures, an important example of a continuous mixture is the scale mixture of normals.

$$p\left(y|\tau\right) = \int \phi\left(y|\mu, \sigma\right) \pi\left(\sigma|\tau\right) d\sigma \qquad (1.0.2)$$

A scale mixture of a normal distribution simply alters the tail behavior of the distribution while leaving the resultant distribution symmetric. Classic examples include the t distribution and double exponential in which the mixing distributions are inverse gamma and exponential, respectively (Andrews and Mallows (1974)). For our purposes, we desire a more general form of mixing which allows the resultant mixture distribution sufficient flexibility to approximate any continuous distribution to some desired degree of accuracy. Scale mixtures do not have sufficient flexibility to capture distributions that depart from normality exhibiting multi-modality and skewness. It is also well-known that most scale mixtures that achieve thick tailed distributions such as the Cauchy or low degree of freedom t distributions also have rather "peaked" densities around the mode of the distribution. It is common to find datasets where the tail behavior is thicker than the normal but the mass of the distribution is concentrated near the mode but with rather broad shoulders (e.g., Tukey's "slash" distribution). Common scale mixtures cannot exhibit this sort of behavior. Most importantly, the scale mixture ideas do not easily translate into the multivariate setting in that there are few distributions on Σ for which analytical results are available (principally the Inverted Wishart distribution).

For these reasons, I will concentrate on finite mixtures of normals. For a finite mixture of normals, the mixing distribution is a discrete distribution which puts mass on K distinct values of μ and Σ.

$$p\left(y|\pi, \{\mu_k, \Sigma_k\}\right) = \sum_k \pi_k \phi\left(y|\mu_k, \Sigma_k\right) \qquad (1.0.3)$$

$\phi(\ )$ is the multivariate normal density.

$$\phi\left(y|\mu, \Sigma\right) = (2\pi)^{-d/2}\, |\Sigma|^{-1/2} \exp\left\{-1/2\left(y - \mu\right)' \Sigma^{-1}\left(y - \mu\right)\right\} \qquad (1.0.4)$$

d is the dimension of the data, y. The K mass points of the finite mixture of normals are often called the *components* of the mixture. The mixture of normals model is very attractive for two reasons: (1) the model applies equally well to univariate and multivariate settings; and (2) the mixture of normals model can achieve great flexibility with only a few components.

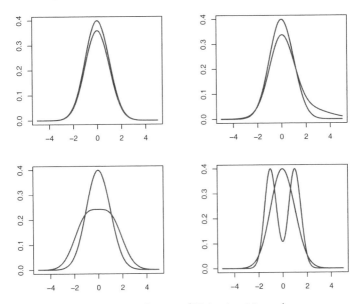

Figure 1.1. Mixtures of Univariate Normals

Figure 1.1 illustrates the flexibility of the mixture of normals model for univariate distributions. The upper left corner of the figure displays a mixture of a standard normal with a normal with the same mean but 100 times the variance (the red density curve), that is the mixture $.95N(0, 1) + .05N(0, 100)$. This mixture model is often used in the statistics literature as a model for outlying observations. Mixtures of normals can also be used to create a skewed distribution by using a "base" normal with another normal that is translated to the right or left depending on the direction of the desired skewness.

The upper right panel of Figure 1.1 displays the mixture, $.75N(0, 1) + .25N(1.5, 2^2)$. This example of constructing a skewed distribution illustrates that mixtures of normals do not have to exhibit "separation" or bimodality. If we position a number of mixture components close together and assign each component similar probabilities, then we can create a mixture distribution with a density that has broad shoulders of the type

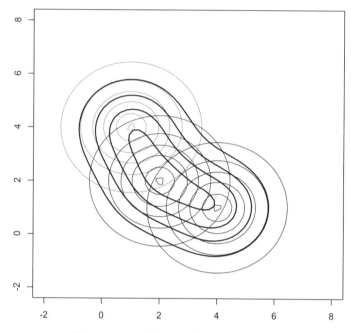

Figure 1.2. A Mixture of Bivariate Normals

displayed in many datasets. The lower left panel of Figure 1.1 shows the mixture $.5N(-1, 1) + .5N(1, 1)$, a distribution that is more or less uniform near the mode. Finally, it is obvious that we can produce multi-modal distributions simply by allocating one component to each desired model. The bottom right panel of the figure shows the mixture $.5N(-1, .5^2) + .5N(1, .5^2)$. The darker lines in Figure 1.1 show a unit normal density for comparison purposes.

In the multivariate case, the possibilities are even broader. For example, we could approximate a bivariate density whose contours are deformed ellipses by positioning two or more bivariate normal mixtures along the principal axis of symmetry. The "axis" of symmetry can be a curve allowing for the creation of a density with "banana" or any other shaped contour. Figure 1.2 shows a mixture of three uncorrelated bivariate normals that have been positioned to obtain "bent" or "banana-shaped" contours.

There is an obvious sense in which the mixture of normals approach, given enough components, can approximate any multivariate density (see Ghosh and Ramamoorthi (2003) for infinite mixtures and Norets and Pelenis (2011) for finite mixtures). As long as the density which is approximated by the mixture of normals damps down to zero before reaching the boundary of the set on which the density is defined, then mixture of normals models can approximate the density. Distributions (such as truncated distributions) with densities that are non-zero at the boundary of the sample space will be problematic for normal mixtures. The intuition for this result is that if we were to use extremely small variance normal components and position these as needed in the support of the density then any density can be approximated to an arbitrary degree of precision with enough normal components. As long as arbitrarily large samples are allowed, then we can afford a larger and larger number of these tiny normal components. However, this is a profligate and very inefficient use of model parameters. The resulting approximations, for any given sample size, can be very non-smooth, particularly if non-Bayesian methods are used. For this reason, the really interesting question is not whether the mixture of normals can be the basis of a non-parametric density estimation procedure, but, rather, if good approximations can be achieved with relative parsimony. Of course, the success of the mixture of normals model in achieving the goal of flexible and relatively parsimonious approximations will depend on the nature of the distributions that need to be approximated. Distributions with densities that are very non-smooth and have tremendous integrated curvature (i.e., lots of wiggles) may require large numbers of normal components.

The success of normal mixture models is also tied to the methods of inference. Given that many multivariate density approximation situations will require a reasonably large number of components and each component will have a very large number of parameters, inference methods that can handle very high dimensional spaces will be required. Moreover, the inference methods that over-fit the data will be particularly problematic for normal mixture models. If an inference procedure is not prone to over-fitting, then inference can be conducted for models with a very large number of components. This will effectively achieve the non-parametric goal of sufficient flexibility without delivering unreasonable estimates. However, an inference method that has no method of curbing over-fitting will have to be modified to penalize for over-parameterized models. This will add another burden to the user—choice and tuning of a penalty function.

1.1 Finite Mixture of Normals Likelihood Function

There are two alternative ways of expressing the likelihood function for the mixture of normals model. This first is simply obtained directly from the form of the mixture of normals density function.

$$L\left(\pi, \{\mu_k, \Sigma_k, k = 1, \ldots, K\} | Y\right) = \prod_i \sum_k \pi_k \phi\left(y_i | \mu_k, \Sigma_k\right)$$

$$(1.1.1)$$

Y is a matrix whose ith row is y_i'. A useful alternative way of expressing the likelihood function is to recall one interpretation of the finite mixture model. For each observation, an indicator variable, z_i, is drawn from a multinomial distribution with K possible outcomes each with probability π_k. y_i is drawn from the multivariate normal component corresponding to outcome of the multinomial indicator variable. That is, to simulate from the mixture of normals model is a two-step procedure:

$$z_i \sim \text{MN}(\pi)$$
$$y_i \sim \text{N}\left(\mu_{z_i}, \Sigma_{z_i}\right)$$

Using this representation we can view the likelihood function as the expected value of the likelihood function given z.

$$L\left(\pi, \{\mu_k, \Sigma_k\}\,|Y\right) = \mathbb{E}\left[L\left(z, \mu_k, \Sigma_k\right)\right]$$

$$= \mathbb{E}\left[\prod_i \sum_k I\left(z_i = k\right)\phi\left(y_i, |\mu_k, \Sigma_k\right)\right]$$

$$(1.1.2)$$

The likelihood function for the finite mixture of normals model has been extensively studied (see, for example, McLachlan and Peel (2000)). There are several unusual features of the mixture of normals likelihood. First, the likelihood has numerous points where the function is not defined with an infinite limit (for lack of a better term, I will call these poles). In a famous example given by Quandt and Ramsey (1978), the likelihood for a mixture of two univariate normals can be driven to any arbitrary value by taking one of the means to be equal to y_i and letting σ for that mixture component go to zero.

$$\lim_{\sigma \to 0} \frac{1}{2\pi\sigma}\exp\left\{-1/2\left(\frac{y_i - \mu_k}{\sigma}\right)^2\right\}|_{\mu_k = y_i} = \infty$$

This means that there are poles for every y value in the data set. Figure 1.3 plots the likelihood function for a mixture of two univariate normals and shows the log-likelihood surface around values of μ close to a particular y_i. This sort of feature may make it difficult for standard optimizers to explore the likelihood surface.

However, it is not poles that present the most difficult problem for exploring the likelihood surface using conventional optimizers that use local derivative information. The mixture of normals likelihood function has $K!$ modes, each of equal height. These modes correspond to all of the possible ways to reorder the labeling of the likelihood normal mixture component parameters. That is, there is no difference between the likelihood

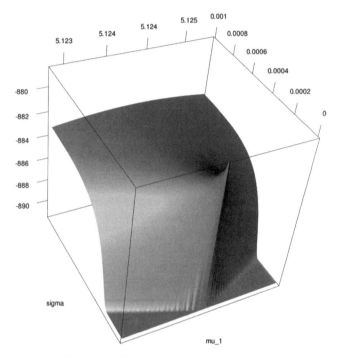

Figure 1.3. Pole in the Likelihood Function

for $\mu_1 = \mu_1^*, \mu_2 = \mu_2^*, \sigma_1 = \sigma_1^*, \sigma_2 = \sigma_2^*$ and the likelihood for $\mu_1 = \mu_2^*, \mu_2 = \mu_1^*, \sigma_1 = \sigma_2^*, \sigma_2 = \sigma_1^*$. Moreover, there are saddle points between these symmetric modes. Figure 1.4 shows what appears to be a saddle point in the likelihood of a mixture of two normals. The likelihood is only depicted in the μ_1, μ_2 space conditional on the values of the standard deviations parameters. The figure shows two local maxima near the points (2,4) and (4,2). However, if you constrain the means to be equal, there is a local maximum at the top of the saddle point near the point (1.5,1.5). This means that any standard optimizer that begins at the point of equal means (not an unreasonable starting point, for example, to start at the mean of the data for all μ parameters) will converge to local maximum that is not the global.

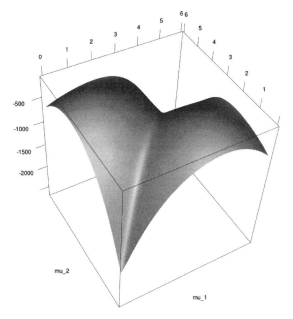

Figure 1.4. Saddle Points in the Likelihood Function

1.2 Maximum Likelihood Estimation

Given the existence of both poles and saddle points, the maximization of the mixture of normals likelihood (1.1.1) is challenging. Standard quasi-Newton optimizers will have to be started from points close to one of the $K!$ global optima in order to work well. If the mixture is designed to approximate a density in high dimensions (5 or more) and a relatively large number of components are desired for flexibility, then the number of parameters in the likelihood can be very large.

$$n_{parm} = K - 1 + K \left(\frac{d(d+1)}{2} \right) + Kd = o\left(d^2\right) \qquad (1.2.1)$$

d is the dimension of the data and K is the number of components in the mixture. The first term in (1.2.1) represents the number of parameters in π, the second from the set of normal component variance-covariance matrices (Σ_k), and the third from the normal component mean vectors. For example, a modest problem with $d = 5$ and $K = 10$ would feature 209 unique parameters. The sheer number of parameters does not present a computational challenge in and of itself. However, the irregular features of the mixture likelihood make high dimensional optimization difficult. Not only are the parameters large in number, but there are constraints on the parameter space imposed by the condition that all of the component covariance matrices must be positive definite symmetric. The nature of these constraints do not facilitate direct input into standard non-linear programming methods which allow for constrained optimization. Instead, it would be advisable to reparameterize to enforce symmetry and positive definiteness.

$$\Sigma_k = U_k' U_k$$

$$U_k = f(\lambda_k, \theta_k) \tag{1.2.2}$$

$$f(\lambda_k, \theta_k) = \begin{bmatrix} e^{\lambda_{k,1}} & \theta_{k,1} & \cdots & \theta_{k,d-1} \\ 0 & e & \ddots & \vdots \\ \vdots & \ddots & \ddots & \theta_{k,(d-1)d/2} \\ 0 & \cdots & 0 & e^{\lambda_{k,d}} \end{bmatrix}$$

However, even with this reparameterization, the mixture of normals likelihood remains a severe challenge for standard optimization methods, particularly for cases where the dimension of the data is greater than one.

1.2.1 E-M Algorithm

Most agree that the only reliable way to compute maximum likelihood estimates for the mixture of normals model is to employ the E-M algorithm. The E-M algorithm is particularly appropriate for those problems which can be characterized as an incomplete data problem. An incomplete data problem is one in which what we observe in our sample can be viewed as a subset of "complete" data. As we have seen, the mixture of normals model can be viewed as a sampling mechanism in which draws are made from a latent (unobserved) indicator vector, z, which indicates which of the K normal components each observation is drawn from. The complete data is then (z, y). The likelihood for the observed data is the complete data likelihood with the unobserved component integrated out. In the mixture of normals case, the integration simply consists of weighting each component by the mixing or multinomial probability and adding the components up. The E-M algorithm is an iterative procedure consisting of an "E-step" and an "M" or maximization step. Given the model parameters, the "E-step" consists of taking the expectation of the unobserved latent indicators and the "M-step" consists of maximizing the component density parameters in the expectation of the complete data likelihood function (see, for example, McLachlan and Peel (2000), section 2.8). As is well-known, the E-M algorithm provides a method by which an improvement (or at least no decrease in the likelihood) can be achieved at each step. This means that the E-M method provides a reliable, if somewhat slow, method of climbing to a local maximum in the mixture of normals likelihood.

The complete data log-likelihood can be written conveniently for application of the E-M method as follows:

$$\log \left(L_c \left(H, \Psi \right) \right) = \sum_{i=1}^{n} \sum_{k=1}^{K} h_{ik} \left(\log \left(\pi_i \right) + \log \phi \left(y_i | \mu_k, \Sigma_k \right) \right)$$

$$(1.2.3)$$

h_{ik} is a matrix of indicator variables. $h_{ik} = 1$ if observation i is from component k. Ψ is the collection of all of the mixture parameters: $\pi, \{\mu_k, \Sigma_k\}$ $k = 1, \ldots, K$. The E-M method starts from an initial value, Ψ^0, of Ψ.

E-Step: Take the expectation of the complete data log-likelihood with respect to the unobserved h_{ik} values. The expectation is taken given Ψ^0.

$$\mathbb{E}\left[\log\left(L_c\left(H,\Psi\right)\right)\right] = \sum_{i=1}^{n}\sum_{k=1}^{K}\mathbb{E}\left[h_{ik}|\Psi^0\right]\left(\log\left(\pi_i\right)\right.$$
$$\left. + \log\phi\left(y_i|\mu_k,\Sigma_k\right)\right) \qquad (1.2.4)$$

$$\mathbb{E}\left[h_{ik}|\Psi^0\right] = \tau_k\left(y_i|\Psi^0\right)$$

$$\tau_k\left(y_i|\Psi^0\right) = \pi_k^0\phi\left(y_i|\mu_k^0,\Sigma_k^0\right)/\sum_{j=1}^{K}\pi_j^0\phi\left(y_i|\mu_j^0,\Sigma_j^0\right) \qquad (1.2.5)$$

M-Step: Maximize the expectation of the log-likelihood computed in 1.2.4 to form new estimates of the component mixture parameters, Ψ^1.

$$max_\Psi Q\left(\Psi|y\right) = \sum_{i=1}^{n}\sum_{k=1}^{K}\tau_k\left(y_i|\Psi^0\right)\left(\log\left(\pi_i\right)+\log\phi\left(y_i|\mu_k,\Sigma_k\right)\right)$$
$$(1.2.6)$$

The solutions to the maximization problem are simply weighted averages of means and covariance matrices:

$$\pi_k^1 = \sum_{i=1}^{n}\tau_k\left(y_i|\Psi^0\right)/n \qquad (1.2.7)$$

$$\mu_k^1 = \sum_{i=1}^{n}\tau_k\left(y_i|\Psi^0\right)y_i/\sum_{i=1}^{n}\tau_k\left(y_i|\Psi^0\right) \qquad (1.2.8)$$

$$\Sigma_k^1 = \sum_{i=1}^{n}\tau_k\left(y_i|\Psi^0\right)\left(y_i-\mu_k^1\right)\left(y_i-\mu_k^1\right)'/\sum_{i=1}^{n}\tau_k\left(y_i|\Psi^0\right)$$
$$(1.2.9)$$

Thus, the E-M method is easy to program and involves only evaluating the normal density to compute the probability[1] that each observation belongs to each of the K components (1.2.5) and simple updates of the mean and covariance matrices.

The E-M method is not complete without advice regarding the choice of a starting point and a method for computing an estimate of the sampling error. Given that the E-M method can be very slow to converge, it is important to choose reasonable starting points. Some advise clustering the data using standard clustering methods and then using the cluster proportions, means, and covariance matrices as a starting point. Regarding the computation of standard errors for the parameter estimate, it appears the most practical approach would be to start a Quasi-Newton optimizer from the last E-M iterate value and use the Hessian estimate to form an approximate information matrix value which can be used for the standard asymptotic normal approximation to the sampling distribution of the MLE.

In many applications, the mixture of normals density approximation will only be one part of a larger model where the emphasis will be on inference for other model parameters. For example, suppose we are to use mixture of normals as the distribution of a regression error term. In that case, our interest is not regarding the error term density parameters but on parameters governing the regression function. Having an MLE procedure (however reliable) for the density parameters is only useful as part of a more general estimation procedure.

Even if our goal is simply to estimate the density of the data, the asymptotic distribution of the MLE for mixture of normal parameters is not directly useful. We will have to compute the asymptotic approximation to the density ordinates.

$$p\left(y|\hat{\Psi}_{MLE}\right) = f\left(\hat{\Psi}_{MLE}|y\right)$$

---

[1]Note that this is the posterior probability of component membership conditional on the normal mixture parameters.

Either the parametric bootstrap or the delta method would be required to obtain an asymptotic approximation to the distribution of the density ordinate and this asymptotic approximation would have to be computed for each potential value of the density ordinate.

Another major problem with a maximum likelihood approach is that the likelihood function will always increase as K increases. This illustrates the "greedy" nature of the MLE approach in which estimators are chosen via minimization of a criterion function (log-likelihood), namely that any increase in flexibility will be rewarded. At its most ridiculous extreme, a mixture of normals that allocates one component to each observation will have the highest likelihood value. In practice, this property of the MLE results in over-fitting. That is, the MLE will attempt to allocate components to tiny subsets of the data in order to fit anomalous values of the data. This propensity of m-estimators for chasing noise in the data is well-known. In order to limit this problem, procedures are used to either "test" for the number of components or to penalize models that offer slight improvements in fit at the expense of many additional parameters. In the mixture of normals literature, both the AIC and BIC criteria have been proposed to help keep the number of components small and to choose among models with differing numbers of components. The BIC criteria has been derived as an approximate Bayes Factor using asymptotic arguments.

In summary, the mixture of normals model provides a formidable challenge to standard inference methods. Even though there is a well-defined likelihood, maximization of this likelihood is far from straightforward. Even abstracting from the practical numerical difficulties in fitting high dimensional mixture of normals models, the problem of over-fitting still must be overcome. Ad hoc criteria for selecting the number of mixture components do not solve the over-fitting problem. What would be desirable is an inference procedure that is numerically reliable, not prone to over-fitting, provides accurate and easy to compute inference methods, and can be easily made a part of a more complicated model structure. The Bayesian methods discussed in the next section of this chapter provide a solution to many of these problems.

1.3 Bayesian Inference for the Mixture of Normals Model

The likelihood function for the mixture of normals presents serious challenges for any estimator based on minimization of a criterion function (such as the MLE). Not only is it difficult to find roots of the score function, but in the normal mixtures problem the parameters, in most cases, do not have a direct meaning. Rather the model is fit with an interest in making inferences regarding an unknown joint density of the data. In a Bayesian setting, the "density estimation" problem is viewed as the problem of computing the predictive distribution of a new value of y. That is, we have an observed data set, Y (the matrix of observations, a $n \times d$ matrix). We assume that these observations are iid draws from a common but unknown distribution. Inferences are desired for the distribution of a y value drawn from the unknown population distribution given the observed data. The predictive density requires that we integrate over the posterior distribution of the unknown parameters, θ.

$$p\,(y|Y) = \int p\,(y|\theta)\,p\,(\theta|Y)\,d\theta \qquad (1.3.1)$$

1.3.1 shows that the parameters are merely devices for implementing a solution to the density estimation-inference problem. What is desired is a method that will allow us to make inferences regarding the posterior distribution of the joint density at any particular value of y. It turns out that mixture of normals model is a model which is particularly well-suited to computationally tractable and accurate approximations to the posterior distribution of density ordinates. Simulation-based methods will be used to navigate the parameter space and avoid the problems associated with derivative-based procedures.

1.4 Priors and the Bayesian Model

A Bayesian analysis of the mixture of normals models starts with a specification of prior distributions for the model parameters. Priors will assume an important role in the analysis of the mixture of normals model as any non-trivial application will involve a very large number of parameters. The standard prior for mixture of normals models is chosen so that the priors for each of the sets of parameters π and $\{\mu_k, \Sigma_k\}$ are conditionally conjugate priors. The choice of conditionally conjugate prior is convenient and exploited by many MCMC methods. However, conjugate priors are not overly restrictive if the source of prior information is some smoothness notion and some views regarding the probable locations of normal components. That is, there is no need to consider priors other than these conjugate priors unless one has very informative views regarding certain component parameters.

The standard priors for the mixture of components model are a Dirichlet distribution for the mixing probabilities and a standard conjugate prior for each of the normal component parameters.

$$\pi \sim \text{Dirichlet}(\alpha) \tag{1.4.1}$$

$$\Sigma_k \sim \text{IW}(v, V) \tag{1.4.2}$$

$$\mu_k | \Sigma_k \sim \text{N}\left(\bar{\mu}, a_\mu^{-1} \Sigma_k\right) \tag{1.4.3}$$

The joint prior for all mixture parameters is obtained by assuming that all parameters $\pi, \{\mu_k, \Sigma_k\}$ are independent. Recall that the Dirichlet is conjugate to the multinomial distribution and the IW[2]-normal prior is conjugate to the multivariate normal distribution. The Dirichlet distribution can be interpreted as a distribution over distributions. That is, a draw from the Dirichlet

---

[2]The IW parameterization we use is given by $p(\Sigma|v, V) \propto |V|^{v/2}|\Sigma|^{-(v+d+1)/2} etr\left(-\frac{1}{2}V\Sigma^{-1}\right)$. Σ is $d \times d$. In this parameterization, $E[\Sigma] = \frac{1}{v-d-1}V$. IW stands for Inverted Wishart.

distribution is a distribution on a discrete probability space. In our case, the draw from the Dirichlet distribution is a particular normal mixture or probability distribution over the K possible components. To extend the Dirichlet distribution to the infinite component Dirichlet process case, it will be useful to define the Dirichlet hyper-parameter as $\alpha = \alpha^* m$ where m is a vector of probabilities that define a base measure.

$$p\left(\pi | \alpha^*, M\right) = \frac{\Gamma\left(\alpha^*\right)}{\prod_{k=1}^K \Gamma\left(\alpha^* m_k\right)} \prod_{k=1}^K \pi_k^{\alpha^* m_k - 1} \qquad (1.4.4)$$

M is the base measure over the discrete probability space of component membership. α^* is a tightness parameter over the base distribution. As α^* increases, the Dirichlet distribution tightens down over the base measure.

One interpretation is that the Dirichlet distribution is the posterior from a previous analysis of multinomial outcome data conducted with a diffuse prior. Under the parameterization given in (1.4.4), α^* can be interpreted as the size of the imaginary sample for which the Dirichlet is the posterior. M is the frequency distribution of multinomial outcomes in this imaginary sample. In this sense, the value of α^* determines the informativeness of the prior – α^* values on par with the size of the sample will constitute a very informative prior. A useful a priori choice of M would be to put equal probability on each component, $m_i = 1/K$. There is also a close connection between infinite mixture models and finite mixture models as I show in section 2.2. These limiting results use the Dirichlet prior parameterization, $\alpha = \frac{\alpha^*}{K}$.

The Dirichlet prior has the effect of putting prior probability mass on finite mixture models with differing numbers of components, in the sense that draws from the Dirichlet distribution can result in mixture probabilities that put a very low probability on models with more than a subset of the K components active. To see this, I compute the prior probability of differing numbers of unique components (that is, by drawing

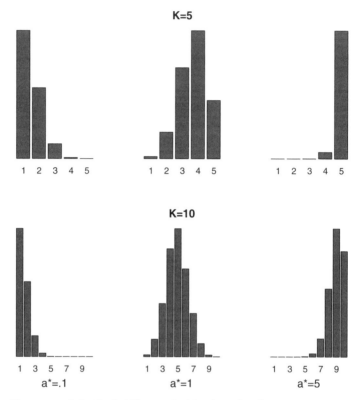

Figure 1.5. Prior Probability on the Number of Unique Components

from the Dirichlet prior and then drawing from the resulting multinomial distributions). Figure 1.5 shows the distribution of the number of unique components for various prior settings. Although the base measure always puts equal probability on all components, the tightness parameter has a dramatic influence on the number of unique components that the Dirichlet prior puts appreciable mass on. Smaller values of α^* (denoted a^* in the figure) are associated with high prior probability on models with only a small number of components out of the total of K possibilities. However, the effect of the tightness parameter is modified by the number of components. For a larger number of components, a given level of α^* will result in high prior probability for a larger number of components. For example, for $\alpha^* = 5$ the prior for a 5 component model has a distribution with 5 as the modal number of active components, while for a 10 component model the prior has 9 as the modal number of components. If there are no practical problems with specifying a model with a relatively large number of components, it may be useful to do so, as for models with a large value of K it is possible to specify a prior which puts substantial prior probability on a wide range from 1 to a moderately large number of components.

It should be emphasized that the Dirichlet prior always puts non-zero probability on any number of unique components up to and including K. However, in situations with very small sample sizes, care should be exercised in specification of the Dirichlet prior. In the literature, there is a tendency to use very small values of α^* on the grounds that this constitutes "vague" or diffuse prior information. This relies on the intuition that we can regard α^* in our parameterization as the size of the pseudo data set used to construct the prior. The analysis here shows that this is a somewhat naive interpretation. Using very small values of α^* can effectively limit the flexibility of the mixture of normals model since this puts very low prior probability on models with more than one component (for small enough α^*).

The priors on the mixture component normal parameters are also important in the specification of the prior. These consist of a standard conjugate prior for the multivariate normal distribution. This Normal-Inverted Wishart prior is jointly conjugate for μ, Σ and will be used in the Gibbs sampler to make a one-for-one draw of these parameters. There are two criticisms that can be made of the conjugate prior. First, the prior on μ depends on the prior on Σ in the way naturally suggested by a strictly data-based prior. Second, the IW distribution has a very heavy right tail and puts prior probability on very large covariance matrices when the degrees of freedom parameter is low (barely above the dimension of the data). In the univariate case, some prefer the log-normal or gamma prior to the IW.

However, in multivariate settings, there is limited freedom in choice of priors unless there is a very efficient method to conduct posterior inference with non-conjugate priors. For example, the Metropolis methods advocated by Daniels and Kass (1999) will not work well in high dimensional (10 or more) problems. Given the computational tractability of the conjugate IW prior, a convincing argument would have to be made that the IW prior, when used in the mixture of normals context, imparts unreasonable prior information which damages the flexibility of the mixture of normals model and should be abandoned in favor of a more reasonable prior. In my experience with mixture of normals models, I have found that care must be taken in the assessment of an IW prior but that the prior, when properly assessed, is very useful.

In the mixture of normals model, the priors on μ, Σ are important to the extent that these priors limit flexibility. For example, a prior that puts a great deal of mass on normal components located close to the origin could limit the flexibility of the normal mixture model to accommodate deviations from normality. It is important, therefore, that the prior on μ allow a sufficient range in the means of the normal components. That is, the role of the μ prior should be to allow for a wide (but not too wide) dispersion of the locations of the normal components. Similarly, a prior that puts mass on only "small" normal components (i.e., those with small variance-covariance matrices) could force the mixture of normals model to use many "small" normal components to approximate the distribution of the data. The prior on Σ should allow for a large range of correlation patterns[3] as well as variability. In other words, the prior on μ, Σ should allow the normal mixture components to be spread across the relevant range of the data as well as to allow for a wide range of "shapes" or covariance matrices.

---

[3] It is common for kernel smoothing methods to use diagonal or even scalar covariance matrices in many implementations for more than one dimension.

In order to assess a reasonable but not overly dogmatic prior, I have found it useful to scale[4] the data so that it has roughly unit variance and is located around the origin. This simplifies the assessment of the prior on μ, Σ. With scaled data, I can then assess the key hyper-parameters of the priors. To simplify assessment and to recognize that the data is centered, I use the following hyper-parameters:

$$\Sigma \sim \text{IW}(\nu, \nu I) \tag{1.4.5}$$

$$\mu | \Sigma \sim \text{N}\left(0, a_\mu^{-1}\Sigma\right) \tag{1.4.6}$$

This parameterization centers the IW prior roughly on the identity matrix, $E[\Sigma] = \frac{\nu}{\nu-d-1}I$. d is the dimension of the data. Small values of ν will assess a prior on Σ with a wide variation. I use a setting of $\nu = d + 3$. This keeps the IW prior proper and avoids the very fat tail associated with a degrees of freedom parameter at $d + 1$. a_μ^{-1} is set to 100 to allow for a very wide range of possible locations for μ. One could argue that this is a too diffuse prior in the sense that it will allow for μ values quite a bit farther than 3 standardized units away from the origin. I term these prior settings the "default" prior.

Figure 1.6 shows draws from the default prior on μ, Σ in terms of the associated normal distributions. The normal components are depicted by contours corresponding to the same level sets for all components. Each component is drawn in a different shade of grey and the μ vector is shown as a solid dot. The contours have a wide variety of shapes and spread. The components are located at means which are spread over a very wide range of possible locations.

The prior settings shown in Figure 1.6 represent a reasonably diffuse prior which spreads prior mass over components of

---

[4]By default, my software scales by demeaning the data and dividing by the appropriate standard deviation. However, one could also center the data using the median and the range. The only point of scaling is to confine the bulk of the data to some sort of cube.

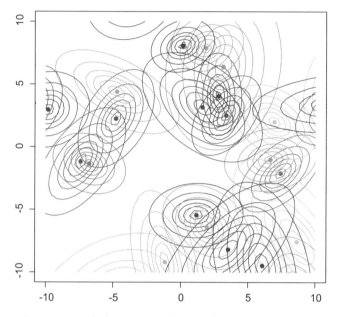

Figure 1.6. Default Prior Distribution of Normal Components

various size, shape, and location. However, even this diffuse prior will result in a posterior that exhibits a good deal of smoothness as there is a natural tendency of any Bayesian procedure with proper priors to only activate components that appreciably improve fit. I will explore this property further later in the chapter. For now, it is useful to ask—what sort of prior settings would induce a posterior that has little smoothing and could behave more like a kernel smoothing method with a small band width.[5] Figure 1.7 shows a different prior settings that

---

[5]Kernel smoothing methods are methods which "smooth" by averaging the empirical distribution function of the data. The bandwidth parameter in kernel smoothing determines how "local" the averaging is. With a large bandwidth, the kernel smoothing density estimate will average over large areas of the data and may miss important features of the data. With a very small bandwidth parameter,

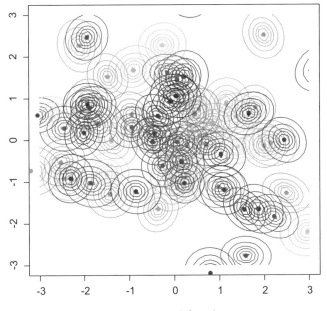

Figure 1.7. A Tight Prior

puts high prior probability on "small" normal components. The settings are $\alpha_\mu^{-1} = .01$, $\nu = d + 50$, $V = .03I$. We can see that the prior draws produce "tight" or small normal components spread across the possible support of the data. Note that the range of possibilities for the locations is somewhat restricted even though the same value of α_μ is used in both. This is because of the link between the size of Σ and the variance of μ in the joint conjugate prior. However, this link does not restrict

---

the kernel methods will produce very non-smooth density estimates. I should note that the "diffuse" prior given by the settings outlined here will not behave like a kernel smoothing method with a large bandwidth. The data will modify the prior and if there is sufficient data with a pattern that requires multiple normal components, these components will be present with non-negligible probability in the posterior. Effectively, the hierarchical models used here can be viewed as an adaptive smoothing method.

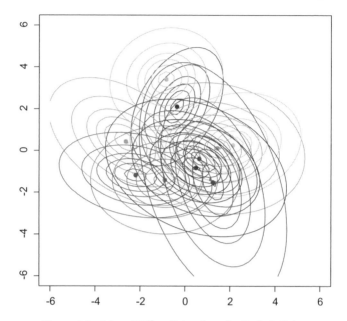

Figure 1.8. A Less Diffuse Prior than the Default Prior

the prior so severely as we cannot achieve the desired high probability allocation on "small" or "tight" normal components that are widely spread. One could argue that the "default" prior settings given above are too diffuse given that the data has been scaled. Perhaps, a more reasonable prior would be one achieved by setting $\alpha_\mu^{-1} = 1, \nu = d + 3, V = \nu I$. The implications of these prior settings for the induce prior on the normal components is depicted in Figure 1.8.

Ultimately, the influence of priors depends on the strength of the information in the data. Obviously, since the priors considered here are not dogmatic (in fact, they are very diffuse), sufficient data information will overwhelm any prior. Some would argue that if the resulting posterior inferences are sensitive to choice of prior hyper-parameters, then the Bayesian approach is problematic. It is easy to specify a prior which is extremely diffuse and also one that is informative in ways that are not anticipated by the user. These extreme prior settings will, almost by definition, have an influence on the posterior. The real question is whether or not there exists a way of assessing priors which achieve desired flexibility while retaining smoothing properties. This is why some fairly deep understanding of the role of prior hyper-parameters is important. Later in this chapter, I will demonstrate that, for a wide variety of applications and with only modest amounts of data, variation in the prior parameters in the range considered here is not influential on the resulting posterior. Only extreme prior settings will materially alter the results. Thus, there is a range of prior setting for which the density approximation is both flexible and smooth and hyper-parameter settings in this range can be assessed using simple guidelines.

1.5 Unconstrained Gibbs Sampler

Bayesian inference for the mixture of normals problem depends on the ability to summarize the posterior distribution. The relevant posterior quantity is the joint predictive density of the data, $p(y|Y)$. In the mixtures of normals model, it is straightforward to construct a MCMC method to navigate the posterior of the mixture parameters by constructing a Markov chain with the posterior as its invariant distribution. Simulations from this chain can be used to construct estimates of the posterior distribution of the predictive density to any desired degree of accuracy.

$$p(y|Y) = \mathbb{E}_{\theta|Y}[p(y|\theta)]$$

This density can be estimated by simply averaging the draws from the chain. The estimation error will decline in $\sqrt{R}$ where R is the number of draws made via the Markov chain

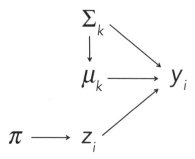

Figure 1.9. Augmented Mixture of Normals Model DAG

simulator.

$$\widehat{p\,(y|Y)} = \frac{1}{R} \sum_{r=1}^{R} \sum_{k=1}^{K} \pi_k^r \phi\left(y|\mu_k^r, \Sigma_k^r\right) \qquad (1.5.1)$$

Formulas for the standard error of averages of correlated series can be used to assess the simulation error and insure that the simulation error is sufficiently small. The superscript, r, represents a draw from the MCMC method.

The mixture of normals model admits a simple unconstrained Gibbs sampler when the parameters are augmented by the indicator vector, z. Each of the n elements of z indicate which of the K components each observation is drawn from. The joint posterior of $z, \pi, \{\mu_k, \Sigma_k\}$ has a marginal posterior that is the same as the posterior of the model formed by combining the mixture of normals likelihood (1.1.1) with the priors introduced in (1.4.3). This observation was first made by Diebolt and Robert (1994). The model with the augmented indicator vectors has a simple Directed Acyclic Graph shown in Figure 1.9.

The Gibbs sampler can be read directly off of the graph in Figure 1.9 as the following set of conditional distributions:

$$\pi \,|\, z, Y \qquad (1.5.2)$$

$$z \,|\, \pi, \{\mu_k, \Sigma_k\}, Y$$

$$\{\mu_k, \Sigma_k\} \,|\, z, Y$$

The key insight behind this Gibbs sampler is that given the indicator vector, z, we have a "classification" of the observations to one of the K components. Given a classification and the fact that we have identical and independent priors for each component and these priors are conjugate priors, then the draws of the normal components parameters are standard Normal-IW draws. The draws of the mixing probabilities given the indicators, $\pi \,|\, z$, are simple draws from the Dirichlet distribution given as the Dirichlet is conjugate to the multinomial distribution. The draws of the indicator function conditional on the mixing probabilities and the normal parameters are multinomials with probabilities given by Bayes theorem for comparison of the K possible normal models.

$$\pi \,|\, z, Y \sim \text{Dirichlet}\,(\tilde{\alpha}) \qquad (1.5.3)$$

$$\tilde{\alpha}_k = n_k + \alpha_k$$

n_k is the number of observations where $z_i = k$. The indicator draws are based on the posterior probability that each observation, y_i, is from the K possible models.

$$z_i \sim \text{MN}(\tilde{\pi}_i)$$

$$\tilde{\pi}_{i,k} = \frac{\pi_k \phi\,(y_i \,|\, \mu_k, \Sigma_k)}{\sum_{j=1}^{K} \pi_j \phi\,(y_i \,|\, \mu_j, \Sigma_j)} \qquad (1.5.4)$$

Inference regarding the normal component parameters follows from standard results for the multivariate regression model (see, for example, Rossi, Allenby, and McCulloch (2005),

Chapter 1).

$$Y_k = \iota\mu_k' + U$$

$$u_i \sim N(0, \Sigma_k)$$

Y_k is the matrix of observations $(n_k \times d)$ corresponding to the kth component. The priors for the component parameters are independent across components with fully conjugate priors.

$$\Sigma_k | Y_k, v, V \sim IW(v + n_k, V + S) \qquad (1.5.5)$$

$$\mu_k | Y_k, \Sigma_k, \bar{\mu}, a_\mu \sim N\left(\tilde{\mu}_k, {}^1\!/\!(n_k + a_\mu)\Sigma_k\right)$$

with

$$S = \left(Y_k - \iota\tilde{\mu}_k'\right)'\left(Y_k - \iota\tilde{\mu}_k'\right) + a_\mu\left(\tilde{\mu}_k - \bar{\mu}_k\right)\left(\tilde{\mu}_k - \bar{\mu}_k\right)'$$

$$\tilde{\mu}_k = \left(n_k + a_\mu\right)^{-1}\left(n_k\bar{y}_k + a_\mu\bar{\mu}_k\right)$$

$$\bar{y}_k = \frac{Y_k'\iota}{n_k}.$$

In the priors here, we set $\bar{\mu}_k = 0$.

The Gibbs sampler is the full-solution to the imputation problem that the E-M problem is trying to solve. Given the classification of observations into normal components, inference is a simple matter. However, we are not certain of the classification of observations and compute the appropriate conditional Bayes Factors $(1.5.4)$[6] to compare the K normal component models. The indicator variables are then drawn with these probabilities. In this manner, the Gibbs sampler does a full accounting regarding the uncertainty of classification of observations. Since the parameters of the normal components are also drawn, the Gibbs sampler effectively integrates out these parameters. The Gibbs

---

[6]Conditional because, in the draw of the indicator vector, we condition on the normal component parameters.

sampler, then, provides a nice intuitive basis for understanding how the posterior distribution in a mixture of normals model is constructed and what kinds of mixture models will have high posterior probability. The Gibbs sampler effectively integrates out the normal component parameters by drawing these from the appropriate conditional posteriors. This illustrates that the posterior distribution effectively uses fully unconditional Bayes Factors to determine the posterior probability of models. The posterior will not put significant mass on models that consist of a very large number of components, each with a small amount of data. This is the fundamental reason that fully Bayesian procedures that use proper priors will tend not to over-fit the data. This is especially important when the mixture of normals model is used in high dimensions and with a large number of components.

The mixture of normals posterior will not put mass on components with very small numbers of observations unless the data assigned to those components differs in a substantial and important way from the rest of the data set. The practical implication of this feature of a Bayesian approach to mixture of normals models is that one can specify a larger number of components and conduct analysis without much concern for over-fitting. Components are added only as necessary to provide flexibility in the density shape. The implication is that methods that develop "tests" for numbers of components are not very useful. Rather than starting with a model with a small number of components and "testing up" by adding more components according to some sort of approximate or exact Bayes Factor, you can start with a large number of components (more than adequate for any clustering or fitting objective) and simply allow components to be shut down in the posterior.

1.6 Label-Switching

As discussed in section 1.1, the mixture of normals likelihood has $K!$ symmetric and equal height modes. The basic unconstrained Gibbs sampler outlined in section 1.5 may "jump" or switch between these modes by simply switching the labeling of the components. That is, instead of the labeling of components $\{\mu_k, \Sigma_k\}$, the sampler can permute the component labels arbitrarily to $\{\mu_{\sigma(k)}, \Sigma_{\sigma(k)}\}$. $\sigma(k)$ defines a permutation of the K indices. A simple example can illustrate this problem. Consider a mixture of two normals, $.5N(1, 1) + .5N(2, 1)$. The means of the two normal components are not sufficiently different to allow for "separation" or a bimodal mixture distribution. In this situation, the Gibbs sampler will tend to switch labels as it navigates the posterior. Intuitively, the sampler may start with the first component associated with a low mean value and the other with a higher mean. However, since the data is not hugely informative regarding the difference in means (that is, there is a non-negligible probability that an observation with value around 1, for example, could have been drawn from the higher mean distribution), the sampler will draw μ_k values from a widely spread conditional posterior. If a high value of μ_1 is drawn, then the sampler may start allocating observations to this as the "high" mean component and, therefore, allocate observations with lower values of y to the second component. At this point in the MCMC sequence, a label switch has occurred. In Figure 1.10, the values of μ_1, μ_2 are displayed in the part of the MCMC draw sequence for this normal mixture with $N = 50$. Several label switches can be observed where the high/low mean label switches.

   The phenomenon of Label-Switching has been discussed extensively as a cause for concern regarding the unconstrained Gibbs sampler. There are a number of approaches to removing Label-Switching: (1) imposition of prior constraints on the component parameters (such as ordinal constraints on the means—here we would impose the constraint $\mu_1 > \mu_2$); (2) various schemes to post-process the draws to "cluster" them (here we would establish some sort of boundary in the means space and label all draws above this boundary as belonging to the "high" mean component and all draws below as belonging to the "low" component); and (3) the permutation

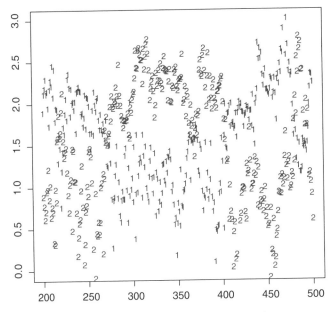

Figure 1.10. Label-Switching Illustrated

sampler of Fruhwirth-Schnatter (2001). In the permutation sampler, the indices of components are randomly permuted on each MCMC iteration, forcing the sampler to navigate all modes.

The first approach to the Label-Switching problem is problematic for two reasons. First, there is a general sense in which a priori restrictions may not be enough to avoid label-switching (see discussion, for example, in Stephens (2000)). There is also a practical issue with multivariate data and multivariate normal component parameters. Restrictions on covariance matrices can destroy conditional conjugacy and can be difficult to formulate. The second problem with the prior restrictions approach is that the restricted Gibbs sampler will, in general, have inferior mixing properties. That is, the information content of a given length simulation run of the restricted Gibbs sampler will be less than that of the unrestricted sampler. The second approach replaces the problem of imposing constraints on the priors with a difficult problem of clustering the unconstrained Gibbs sampler draws. The clustering problem has been proposed as a numerical optimization problem. This problem, in and of itself, may be equally as challenging of a numerical problem as the fitting of the mixture of normals model in the first place.

A "solution" to the Label-Switching problem may be required only for two possible reasons. The first reason is that there is a desire to make inferences about component parameters. This is generally associated with a "clustering" view of the mixture of normals problem. The clustering view is a literal interpretation of the indicator model. Each observation is associated with a specific cluster and the goal of inference is to identify those clusters and classify the observations. In this situation, it will be useful to describe the clusters in terms of their means and covariances. The second reason is that there are some who wish to compute Bayes Factors for comparison of mixtures of normal models with varying numbers of components. Those Bayes Factor methods that rely on posterior simulators will have to be adjusted for the fact that the likelihood has $K!$ modes.

An alternative view is that the mixture of normals model is a way of generating a flexible form for the density of all observations. The latent indicator is just a device for conducting inference and the fundamental posterior of interest is the "marginal" posterior with the indicators integrated out. In this view, component parameters have no meaning and are not necessarily associated with any meaningful clustering of observations. The components are useful as part of an approximation of a joint density. For example, implicit in the clustering view is that the clusters are associated with relatively "small" or, at least, non-overlapping normal components. If there is one "large" or high variance component that covers most of the data range, then this component can not be meaningfully

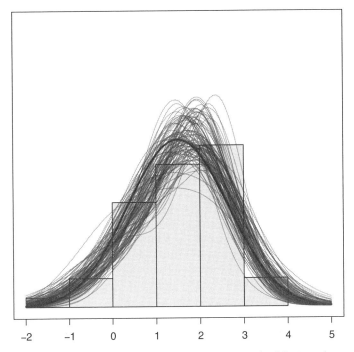

Figure 1.11. Density Draws from the Unconstrained Gibbs Sampler

associated with a cluster. The more compelling interpretation is that this "large" component lays down a base which is modified by the smaller components for the purpose of density approximation.

In this alternative or "marginal" view, the fundamental object of interest is the joint density of the observations. The unconstrained Gibbs sampler provides draws from the posterior of this density. Any density estimate (such as the posterior mean) is label-invariant. That is, permutations of the labels do not affect posterior distribution of density ordinates. Figure 1.11 shows the mixture of two univariate normals discussed above. The thick dark line is the true density and the lighter lines are each density draw. Behind the density curves is the histogram of the data. The draws shown correspond to the same draws of component means displayed in Figure 1.10. As expected, the density draws are label-invariant.

It has become popular to use the mixture of normals model for "model-based" clustering of observations. In my view, this is only appropriate if there are a priori reasons to believe that there are "small" or non-overlapping components associated with groups of observations. In this case, the draws of the indicator vector can be used to infer cluster membership (see section 1.8 for further discussion). Component membership is also label-invariant. Thus, clustering can be based on the output of the unrestricted Gibbs sampler. However, substantive interpretation of the meaning of each cluster will have to be based on other analyses and cannot be obtained solely from the unrestricted Gibbs sampler output.

1.7 Examples

In this section, I will illustrate the effectiveness of the Bayesian approach to mixture of normals with a set of simulated examples. These examples are designed to explore the capability of the mixture of normals approach to approximate non-nested models, i.e., distributions outside the mixture of normal class. In addition, I will explore the performance of the approach in high dimensional problems. Prior sensitivity will also be examined. I will compare the results to the other principal general purpose density approximation method, kernel smoothing.

I will start with an example of a nested application, that is, data will be simulated from a known mixture of normals model. This example will illustrate the rapid mixing properties of the MCMC method as well as the challenges associated with summarizing the results of high dimension density inference. I will simulate 1000 observations from the following five-dimensional mixture of three normal components.

$$y \sim \sum_{k=1}^{3} \pi_k \phi\left(y|\mu_k, \Sigma_k\right) \qquad (1.7.1)$$

$$\mu_1 = \begin{bmatrix} 1 \\ 2 \\ 3 \\ 4 \\ 5 \end{bmatrix} \quad \mu_2 = 2 \times \begin{bmatrix} 1 \\ 2 \\ 3 \\ 4 \\ 5 \end{bmatrix} \quad \mu_3 = 3 \times \begin{bmatrix} 1 \\ 2 \\ 3 \\ 4 \\ 5 \end{bmatrix}$$

$$\Sigma_1 = \Sigma_2 = \Sigma_3 = \begin{bmatrix} 1 & .5 & \cdots & .5 \\ .5 & 1 & \ddots & \vdots \\ \vdots & \ddots & \ddots & .5 \\ .5 & \cdots & .5 & 1 \end{bmatrix}$$

$$\pi = \begin{bmatrix} 1/2 & 1/3 & 1/6 \end{bmatrix}$$

This is a mixture model that is specified so that the separation in the marginal distribution increases from the first to the fifth dimension. Bayesian inference for this model depends on both the number of components specified as well as the prior settings for the Dirichlet and conjugate regressions priors. I will use the "default" prior settings developed in section 1.4, $K = 10$ and $\alpha = (\frac{5}{K})\iota$, $\nu = 5 + 3$, $V = \nu I$. By specifying ten mixture components, the posterior is computed over a total of 209 parameters. The sampler is started from an equal allocation of observations to the ten possible components. Figure 1.12 shows the marginal density of the fifth dimension of the data, y_5, for the first 50 draws from the sampler. The dark thick line is the true marginal density of fifth dimension. The MCMC draws

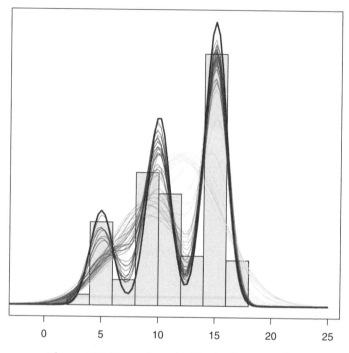

Figure 1.12. Draws from the Marginal Density of y_5

are shown by curves that transform from light to dark as the draw sequence number increases. The first draws are basically a normal density with location given by the overall mean of the data. As the sampler proceeds, the draws very rapidly adapt into a tri-modal distribution. By the twenty-fifth draw or so, the draws have settled into an equilibrium distribution that captures posterior variation around a very close approximation to the true density (note that the data is represented by the light grey histogram).

Figure 1.12 shows that, in this example, the sampler converges extremely rapidly even though the object is a five-dimensional density and the number of mixture component

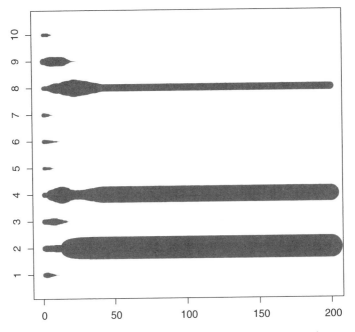

Figure 1.13. Trace of Draw by Draw Component Frequencies

parameters is over 200. It is also instructive to see if the sampler is able to discern that the observed data is from a three component mixture. If our intuition is correct, the natural Bayes Factor computations that allow the sampler to navigate the posterior should favor models with a small number of components. Figure 1.13 shows that, indeed, the MCMC sampler shuts down many of the 10 components and visits the remaining 3 components with frequency very closely approximating the π values. The figure also shows very little evidence of Label-switching. That is, the sampler settles in quickly in the neighborhood of one of the modes and remains in this area because of the concentration of the likelihood near the mode. The Gibbs sampler will tend not to navigate to other modes due to the lack of mass in the posterior in the regions between modes. However, as all

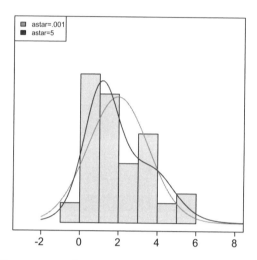

Figure 1.14. Prior Sensitivity—A Univariate Example

modes have equivalent information with respect to the density of observations, there is no loss of information by navigating only a subset of the modes.

In order to investigate the sensitivity of the Bayes procedure to prior settings, I will first start with a univariate example and an extreme case with very little sample information. Consider a mixture of two normals, $.75N(1, 1) + .25N(4, 1)$ with $N = 50$. Here the choice of prior parameters should be expected to influence the posterior distribution. Very small values of the Dirichlet prior parameter, α^*, put a great deal of prior probability on models with a very small number of prior components. Figure 1.14 compares the Bayes estimate of the density of the observations for $\alpha^* = .001$ and $\alpha^* = 5$, with $K = 5$. The fitted density is much more like a one component model for the very small value of the Dirichlet parameter. With a somewhat larger value, the fitted density fits the data more closely as might be expected. Thus, very small values of α^* are too informative and should not be used as these values will destroy the flexibility of the mixture of normals approach.

The density estimates with the default prior are shown by the darker line in Figure 1.14. In spite of the small number of observations and the relatively large number of components, the Bayes estimate is remarkably smooth.

We have seen that Bayesian inference in the normal mixture model quickly adapts and we learn that only a subset of the components are needed to accommodate the data structure. However, this example is not informative with respect to the practical ability of the mixture of normals to approximate densities that cannot be written as normal mixtures. As the number of observations tends to infinity, one has the luxury of using more and more normal components to approximate the distribution of the data . However, this asymptotic property sheds little light on the practical utility of the mixture of normals model. For this reason, I will now consider non-nested models. Three examples of increasing difficulty will be used to gauge the practical value of the mixture of normals model. The sensitivity of the results to prior settings will also be explored.

The first non-nested example will be data drawn from a univariate χ_6^2 distribution. This distribution is quite skewed. Figure 1.15 plots the fitted density from the mixture of normals approach with default prior settings on the normal component parameters. A maximum of five components are used and the Dirichlet prior parameter α^* (recall that I use a parameterization of $\alpha = \frac{\alpha^*}{K}$) is varied. On the top panel, a very small sample size of 100 observations is used and α^* is varied from .1 to 10. The posterior mean is displayed as the lighter line. The darker line is the true density. The black line is a standard kernel smoother implemented in the R function `density`. The densities are plotted on top of a light grey histogram of the data. The Bayes estimates are very smooth and not affected by the choice of prior Dirichlet parameter even though the sample size is rather small. The kernel smoother with optimal bandwidth is much less smooth and does not match well the tail behavior of the true density. With a sample size of 500, the differences between the various methods is minimal.

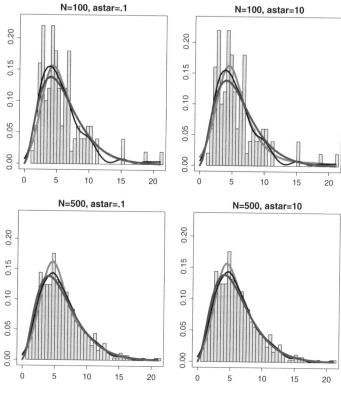

Figure 1.15. Mixture of Normals Fit to χ_6^2

A second example is the distribution discussed by Gelman and Meng (1991). This distribution has normal conditional distributions but the joint is not normal. For some parameter values, the distribution can display contours which look like deformed ellipses and, hence, it is sometimes described as the "banana" distribution. It seems clear that normal components can be positioned along the principal curve of symmetry of the distribution to create the desired shape. The bivariate density and associated conditional densities have the following expressions.

$$p\left(y_1, y_2\right) \propto \exp\left\{-\frac{1}{2}\left(A\left(y_1^2 y_2^2 + y_1^2 + y_1^2 - 2By_1 y_2\right.\right.\right.$$

$$\left.\left.\left. -2C_1 y_1 - 2C_2 y_2\right)\right)\right\}$$

$$p\left(y_1 | y_2\right) \sim N\left(\frac{By_2 + C_1}{Ay_2^2 + 1}, \frac{1}{A\left(y_2^2 + 1\right)}\right) \qquad (1.7.2)$$

$$p\left(y_2 | y_1\right) \sim N\left(\frac{By_1 + C_2}{Ay_1^2 + 1}, \frac{1}{A\left(y_1^2 + 1\right)}\right) \qquad (1.7.3)$$

We can create a sample from this density by using the conditional distributions in (1.7.2–1.7.3) to specify a Gibbs sampler and thin every 100 draws to create what appears to be an iid sample from the banana density. In our example, we set $N = 1000$, $A = .5$, $B = 0$, $C_1 = C_2 = 3$. Figure 1.16 shows the contours of the true density in the upper panel. In the lower panel, contours of the true density and the posterior mean of the mixture of normals approximation are compared. The mixture of normals approximation is very accurate with only minor variations from the true density.

Even with modest sized data sets, the mixture of normals approach appears to well-approximate non-nested distributions. I have not yet examined skewed multivariate distributions. Since kernel smoothing methods cannot be practically extended beyond one or two dimensions, I start with a bivariate log-normal distribution for the purposes of illustrating the difference between kernel smoothing and the Bayesian approach to mixtures of normals.

$$y \sim e^z$$

$$z \sim N\left(0, \begin{bmatrix} 1 & .8 \\ .8 & 1 \end{bmatrix}\right) \qquad (1.7.4)$$

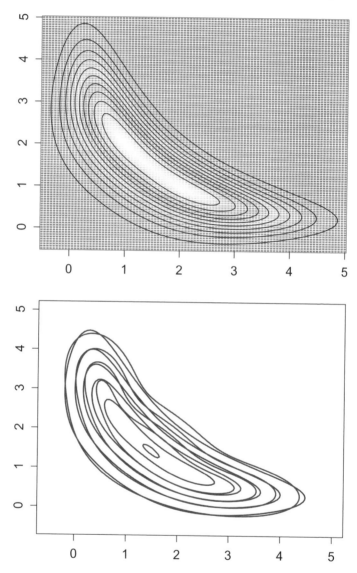

Figure 1.16. Approximating the "Banana" Distribution

N=500, dim(y)=2

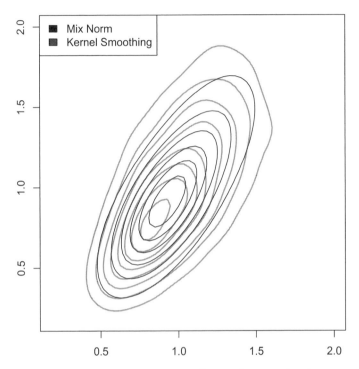

Figure 1.17. Kernel Vs. Mixture of Normals Approximations

Figure 1.17 shows a comparison for bivariate log-normal data and a modest number of observations. The kernel-based contours are grey and the contours based on the posterior mean from the mixture of normals model are shown in black. A maximum of 20 normal components are specified with default prior settings. The kernel-density method produces non-smooth contours. In spite of large number of normal components, the mixtures of normal density estimate is very smooth.

I now consider a five-dimensional log-normal distribution with an equi-correlated covariance matrix with off-diagonal

elements of .8 as in (1.7.4). There are no five-dimensional
kernel smoothing implementations available for R. Figure 1.18
displays a representative (due to the symmetry of the specified
log-normal distribution, all bivariate marginals are identical)
bivariate marginal of the true and Bayes estimate of the log-
normal density for a modest amount of data, $N = 1250$. The top
half shows the prior Dirichlet parameter set to a very small value,
$\alpha = \frac{.1}{K}$, and the bottom displays results for a much larger value,
$\alpha = \frac{10}{K}$. $K = 20$. The contours of the fitted density are shown in
black along with the true density shown in grey. The data points
are also displayed.

In spite of the relatively small number of observations (250
per dimension), the mixture of normals provides a good approximation to the log-normal. The mixture of normals has a total of
419 parameters and yet there is no evidence of over-fitting. The
only deficiency of the approximation is that the fitted density
curves are not as sharp or "pinched" at the origin as the true
log-normal density.[7] The Bayes procedure shows virtually no
sensitivity to the Dirichlet prior parameter, α^*.

The lack of over-fitting by the Bayesian procedure can be
shown by comparing the Bayes estimates of bivariate marginal
distributions for models with different maximal numbers of
normal components. I consider comparisons of a 5 and 20
component model. In order to stress-test the Bayesian procedures, I consider an even smaller sample size of $N = 500$. Many
statisticians would regard the idea of fitting a five-dimensional
distribution with only 500 observations as a somewhat risky
proposition in that this amount of data invites the possibility
of over-fitting. Figure 1.19 compares the contours from 5 and
20 component fits. The 5 and 20 component contours overlap
almost exactly. This illustrates the dramatic sense in which the

---

[7]In Chapter 2, we show that the finite mixture approximation to the log-
normal can be improved upon using hyper-parameter settings suggested by a
infinite mixture Dirichlet process approach. The default prior settings used here
do not put sufficient mass on small components. Priors are placed on the hyper-
parameters to increase flexibility and improve fit. See Figure 2.11.

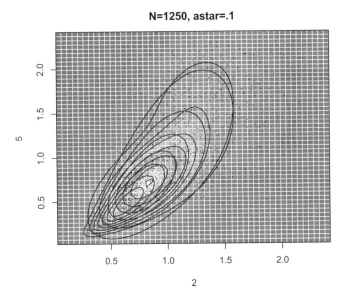

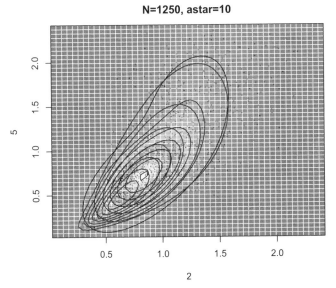

Figure 1.18. Bivariate Marginals from a Five-dimensional Log-normal

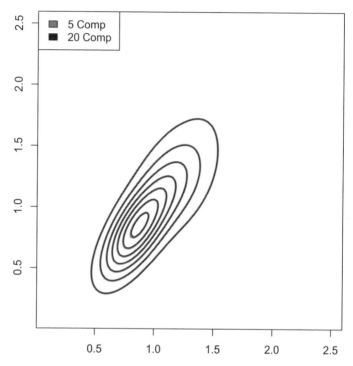

Figure 1.19. Comparison of 5 and 20 Component Fits

Bayes procedure resists over-fitting. As indicated above, this property of the Bayes procedure results from the use of proper priors and the implicit Bayes Factor computations undertaken in exploring the posterior.

1.8 Clustering Observations

In many marketing applications, some sort of clustering method is desired to group "observations" or units that exhibit some sort of similarity. Clustering can be done either on the basis of observed data for each unit or on the basis of model parameters.

Most clustering methods are based on distance metrics that are related to the normal distribution. If a mixture of normals is used, a very general clustering method can be developed. "Observations" are grouped into various normal sub-populations. The only caveat is that there is no restriction that variance of each normal mixture component be "smaller" than the variation across components. For example, a thick-tailed distribution across units can be approximated by the mixture of a small variance normal with a very large variance normal. Observations that get clustered into the large variance component should be interpreted as "similar" only to the extent that they are outliers.

In addition to the metric by which similarity is gauged, there is also a question as to what variables should be used as the basis of clustering. Traditional methods cluster units on the basis of observables such as psycho-graphics. However, if unit level behavioral data is available, it is now possible to cluster units on the basis of unit-level parameters. In some applications, these parameters can be interpreted as representing unit level tastes. This form of clustering on the basis of inference about the unit-level parameters, θ_h, can be termed "behavioral" clustering. Given that psycho-graphics are not very predictive of brand preferences or sensitivity to marketing variables, it is likely that behavioral clustering will be very useful in marketing applications.

To cluster based on a mixture of normals model,[8] we use the latent indicators of component "assignment." We note that this can apply either when a mixture of normals approach is applied directly to data (density estimation) or as the random coefficient distribution. All that is required is that we have draws from the posterior distribution of the indicator variables. These draws of

---

[8] All of the ideas in this section can be applied to cluster models with a fixed or variable number of normal components. The DP process models considered in Chapter 2 will also produce component membership draws which can be clustered using these methods.

the indicator variables, z, can be used to form a similarity matrix

$$S = [s_{i,j}] \tag{1.8.1}$$

$$s_{i,j} = \begin{cases} 0 & if \ z_i \neq z_j \\ 1 & if \ z_i = z_j \end{cases}$$

We note that the similarity matrix is invariant to Label-Switching. (1.8.1) defines a function from a given partition or classification of the observations to the similarity matrix. To emphasize this dependence, we will denote this function as $S(z)$. That is, for any clustering of the observations defined in an indicator vector, we can compute the associated similarity matrix. We can also find, for any similarity matrix, a classification or indicator vector consistent with the given similarity matrix. This function we denote by $z = g(S)$.

By simply averaging over the draws from the marginal posterior of the indicator variables, we can estimate the posterior expectation of the similarity matrix.

$$S^* = \mathbb{E}_{z|data}[S(z)]$$

$$\hat{S^*} = \frac{1}{R} \sum_{r=1}^{R} S(z) \tag{1.8.2}$$

Given the expected similarity matrix, the clustering problem involves the assignment or partition of the units so as to minimize some sort of loss function. Let z be an assignment of units to groups and $L(S^*, S(z))$ be a loss function, then we can define the clustering algorithm as the solution to the following problem:

$$\min_z L\left(S^*, S(z)\right) \tag{1.8.3}$$

In general, this problem is a difficult optimization problem involving non-continuous choice variables. One could adopt two heuristics for the solution of the problem: (1) simply "classify" two observations as in the same group if the posterior expectation of similarity is greater than a cut-off value; and (2) find the posterior draw of z which minimizes loss. A simple loss function would be the sum of the absolute values of the differences between estimated posterior similarity and the implied similarity for a given value of the indicator or classification variable.

$$z_{opt} = \mathrm{argmin}_{\{z'\}} \left[\sum_i \sum_j \left| \hat{S}_{ij} - S(z')[i,j] \right| \right] \qquad (1.8.4)$$

The second heuristic uses the MCMC chain as a stochastic search process. The *bayesm* routine, clusterMix, implements both heuristics.

1.9 Marginalized Samplers

The unconstrained Gibbs sampler in (1.5.2) is a set of three distinct "blocks": (1) draws of the mixture probability given the indicator variables; (2) draws of the indicator variable given the mixture probabilities and normal component parameters; and (3) draws of the normal component parameters given the indicator variable. All three distributions exploit the conditional conjugate priors which facilitate direct draws from the conditional posterior. It is possible, however, to marginalize or collapse this Gibbs sampler by integrating out the normal component parameters. The term "collapsed" Gibbs sampler was coined by Liu (1994) to apply to any situation in which a component is "marginalized" or integrated out of a Gibbs sampling procedure. For example, consider a general setting with parameter vector θ. If θ is partitioned into three components then the standard Gibbs sampler is defined by

$$\theta_1 | \theta_2, \theta_3 \qquad (1.9.1)$$

$$\theta_2 | \theta_1, \theta_3 \qquad (1.9.2)$$

$$\theta_3 | \theta_1, \theta_2. \qquad (1.9.3)$$

If the third component, θ_3, is integrated out from (1.9.1) and (1.9.2), then a collapsed Gibbs sampler is defined over the two components θ_1 and θ_2.

$$\theta_1|\theta_2 \qquad\qquad (1.9.4)$$

$$\theta_2|\theta_1$$

In the mixture of normals model, the collapsed Gibbs sampler would consist only of draws of the indicator vector and the mixture probabilities.

$$\pi|z, Y \qquad\qquad (1.9.5)$$

$$z|\pi, Y$$

Intuitively, we might expect the collapsed Gibbs sampler to have a higher information content or better mixing properties since a set of parameters has been integrated out and we have reduced the dimension of the problem. In the case of the mixture of normals model, this dimension reduction can be very substantial given the large number of normal component mixture parameters. Liu (1994) and Liu, Wong, and Kong (1994) consider the general properties of the collapsed Gibbs sampler and provide some general theory regarding the properties of the full and collapsed Gibbs sampler. Liu, Wong, and Kong (1994) conclude that there is a general sense in which the covariance structure of the collapsed Gibbs sampler can be superior to that of the full sampler (see theorem 5.1, p. 37). A fully satisfactory comparison of the collapsed and full Gibbs samplers would conclude that

$$Cov\left(f\left(\tilde{\theta}_1^r, \tilde{\theta}_2^r\right), f\left(\tilde{\theta}_1^{r-1}, \tilde{\theta}_2^{r-1}\right)\right) \leq Cov\left(f\left(\theta_1^r, \theta_2^r\right),\right.$$
$$\left. f\left(\theta_1^{r-1}, \theta_2^{r-1}\right)\right) \qquad\qquad (1.9.6)$$

where the $\tilde{\theta}^r$ represents draws from the collapsed sampler and θ^r are draws from the full Gibbs sampler. If this inequality held[9]

---

[9]Note that Liu, Wong, and Kong (1994) only consider first order autocorrelations. A complete analysis would be to compute the effective sample size

for any function f then it would be appropriate to conclude that there is a strong theoretical basis for favoring a collapsed Gibbs sampler over the full sampler. However, the results in Liu, Wong, and Kong (1994) are considerably weaker than this. Their results are that the inequalities only apply in a minimax sense (p. 38)—the "worst"[10] function under the collapsed and full Gibbs sampler obeys the inequality in (1.9.6). Therefore, it remains to be determined whether or not the full sampler has higher autocorrelation than the collapsed sampler in a particular model and for particular functions.

In this section, we discuss implementation of the collapsed Gibbs sampler for the mixture of normals model and investigate its autocorrelation properties for some of the examples considered here. The collapsed Gibbs sampler relies on the fact that it is possible to integrate out the normal component parameters and draw the indicator vector, z, directly as in (1.9.5). In particular, the collapsed Gibbs sampler affects a draw of z by drawing successively from predictive distributions for each element of the vector given all of the other

$$p\left(z_j | z_{-j}, \pi, Y_{-j}\right) = \text{MN}(\lambda) \; j = 1, \dots, n \qquad (1.9.7)$$

$$\lambda_k = \frac{\pi_k p\left(y_j | M_k, z_{-j}, Y_{-j}\right)}{\sum_{l=1}^{K} \pi_l p\left(y_j | M_l, Y_{-j}\right)}. \qquad (1.9.8)$$

Here M_k means that we are conditioning on the kth component model. The multinomial distribution of each element of the indicator vector, given all of the others, is based on applying Bayes theorem to the comparison of the k possible models as in the full Gibbs sampler. However, the density of the observations is no longer conditioned on the normal component parameters,

---

from a numerical standard error computed using the entire autocorrelation function.

[10]"Worst" meaning the worst-case or the function which produces the maximum autocorrelation.

but, rather, is based on the predictive distribution.

$$p\left(y_j|M_k, Y_{-j}, z_{-j}\right) = \int p\left(y_j|\mu_k, \Sigma_k\right) p\left(\mu_k, \Sigma_k|z_{-j}, Y_{-j}\right)$$

$$\times d\mu_k d\Sigma_k \qquad (1.9.9)$$

The conditional posterior of the normal component parameters is based on the sufficient statistics associated with the assignment of the observations in Y_{-j} using the component memberships in z_{-j}. As is well-known, the form of this predictive distribution is a multi-variate student t (see, for example, Gelman, Carlin, Stern, and Rubin (2004), p. 88). In particular, for the conjugate priors on μ, Σ in (1.4.3),

$$p\left(y_j|M_k, z_{-j}, Y_{-j}\right) = p\left(y_j|M_k, S_{-j}^k\right) \sim \mathrm{Mvst}\left(v^*, \tilde{\mu}, V^*|S_{-j}^k\right)$$

$$v^* = v_k - d + 1$$

$$\tilde{\mu} = \left(n_k \bar{y}_k + a_\mu \bar{\mu}\right) / \left(n_k + a_\mu\right)$$

$$V^* = \frac{a_\mu^k + 1}{a_\mu^k} V$$

$$V = V + SS_k + \left(\frac{n_k a_\mu}{n_k + a_\mu}\right)(y_k - \bar{\mu})(y_k - \bar{\mu})'$$

$$v_k = v + n_k$$

$$a_\mu^k = a_\mu + n_k \qquad (1.9.10)$$

d is the dimension of the data. S_{-j}^k are the sufficient statistics for the kth normal component based on those observations in Y_{-j} which are associated with model k.

$$S_{-j}^k = \left\{ n_k, \bar{y}_k, SS_k = \sum_{i=1}^{n_k} \left(y_i^{(k)} - \bar{y}_k\right)\left(y_i^{(k)} - \bar{y}_k\right)' \right\} \qquad (1.9.11)$$

$y_i^{(k)}$ are the observations in Y_{-j} which are associated with the kth component. Note that if no observations are allocated to a particular normal component, the posterior predictive density simply becomes the prior predictive density.

To compute the probabilities given in (1.9.7), it is convenient to evaluate the log probabilities and then normalize. The log of the predictive density can be simplified as follows.

$$\log p\left(y_j | M_k, S^k_{-j}\right) = -\frac{d}{2}\log(\pi) + \frac{d}{2}\log\left(\frac{a^k_\mu}{a^k_\mu + 1}\right)$$

$$+ \log\left(\Gamma\left(\frac{v_k + 1}{2}\right)\right)$$

$$- \log\left(\Gamma\left(\frac{v_k - d + 1}{2}\right)\right) + \log(|RI|)$$

$$+ \frac{v_k + 1}{2}\log\left(1 + v'v\right) \tag{1.9.12}$$

$$v = \sqrt{\frac{a^k_\mu}{a^k_\mu + 1}} RI'\left(y_j - \tilde{\mu}_k\right)$$

$$V^{-1} = RI\,RI'$$

RI is the inverse of the Cholesky root of V. Γ is the gamma function. In order to proceed by "Gibbsing" through the indicator vector, the sufficient statistics for each normal component have to be updated. As the sampler proceeds from the jth to the $(j + 1)st$ observation, we need to remove the $(j + 1)st$ observation from the sufficient statistics and add in y_j to the sufficient statistics associated with whatever component is now associated with this observation (the draw of z_j). The updating formulas for adding and subtracting an observation to the sufficient statistics are given below.

$$\bar{y}_{(-i)} = \frac{n\bar{y} - y_i}{n - 1} \tag{1.9.13}$$

$$SS_{(-i)} = SS - y_i y'_i + n\bar{y}\bar{y}' - (n-1)\bar{y}_{(-i)}\bar{y}'_{(-i)}$$

$$\bar{y}_{(+i)} = \frac{n\bar{y} + y_i}{n + 1}$$

$$SS_{(+i)} = \bar{y}_{(+i)} = SS + y_i y'_i + n\bar{y}\bar{y}' - (n+1)\bar{y}_{(+i)}\bar{y}'_{(+i)}$$

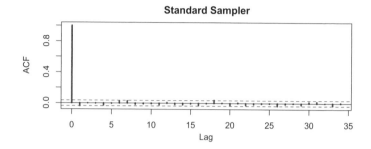

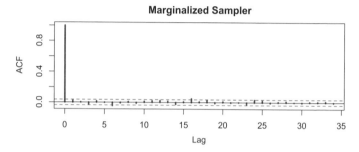

Figure 1.20. Autocorrelation Functions for the Mean: Log-normal Case

   The code to implement the marginalized or collapsed Gibbs
sampler for the mixture of normals model will require a loop
over the entire indicator vector. This is unlike the full Gibbs
sampler where the entire indicator vector is drawn at once and
all of the relevant normal densities can be evaluated in an $n \times K$
matrix. This means that the code for the collapsed Gibbs sampler
will be much slower than the full sampler when the sampler is
implemented in an interpreted language such as R. However,
even implementations in lower level C code are still much
slower than the full sampler. It should also be emphasized that,
in order to implement full Bayes inference regarding the density
of the data (i.e., compute the predictive density in (1.3.1)), we
must draw the component parameters for each indicator draw.
This can be done either alongside of the draws in the collapsed
Gibbs sampler code or a form of post-processing for each of the
indicator vector draws.

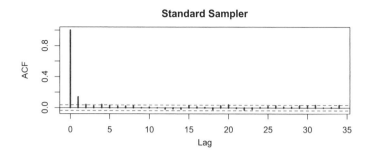

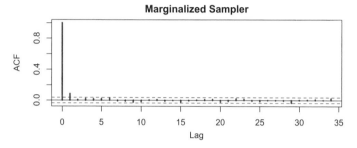

Figure 1.21. Autocorrelation Functions for the Density Ordinate: Lognormal Case

I will now compare the collapsed or marginalized Gibbs samplers when applied to challenging density inference situations. I will return first to the five-dimensional log-normal example. In order to compare the autocorrelation properties of the two samplers, a functional of the density must be chosen. I will use two functions which are of importance in applications of mixture modeling. The first function is the implied first moment of the mixture of normals, $\text{mom}_1^r = \sum_{k=1}^{K} \pi_k^r \mu_k^r$. Both samplers will be used to construct draw sequences of the mean of the normal mixture. The second function is the ordinate of the mixture density evaluated at a specific point. Figure 1.20 shows the autocorrelation functions of the sequence of draws of the mean. The figure shows remarkably low autocorrelation in the draws. Given this low autocorrelation, there are

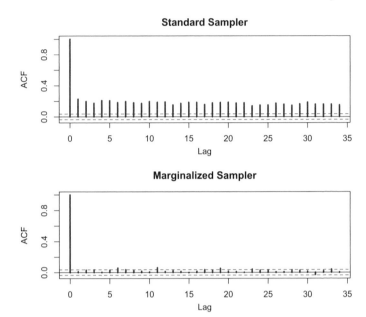

Figure 1.22. Autocorrelation Functions for the Mean: Outlier Example

little differences between the marginalized (collapsed) and full Gibbs samplers. Figure 1.21 displays the autocorrelation density evaluated at the point $(1,1,1,1,1)$ for the full and marginalized samplers. The marginalized sampler shows slightly less autocorrelation.

The next example I consider is a normal mixture model with an outlier component. In outlier examples, there are, by definition, relatively small groups of observations associated with a specific normal component. This can reduce mixing of the chain as once observations are allocated to a "outlier" component it is difficult to break the association between these observations from the "outlier" component. In this example, I generated data from a mixture of three bivariate normal components. One of the components has a very high variance

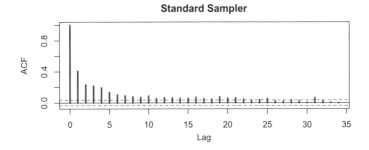

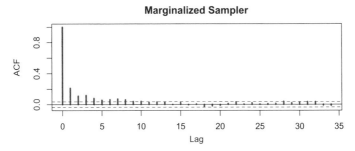

Figure 1.23. Autocorrelation Functions for the Density Ordinate: Outlier Example

and is a standard approach to modeling outliers.

$$\pi = (.475, .475, .05)$$

$$\mu_1 = \begin{bmatrix} 1 \\ 1 \end{bmatrix} \mu_2 = \begin{bmatrix} 4 \\ 1 \end{bmatrix} \mu_3 = \begin{bmatrix} 2.5 \\ 1 \end{bmatrix}$$

$$\Sigma_1 = \Sigma_2 = \begin{bmatrix} 1 & .7 \\ .7 & 1 \end{bmatrix}$$

$$\Sigma_3 = 100 \begin{bmatrix} 1 & 0 \\ 0 & 1 \end{bmatrix}$$

Figure 1.22 displays the autocorrelation of the draws of the mean of the mixture of these three normals using both marginalized and full Gibbs samplers. As expected, the full Gibbs sampler exhibits some autocorrelation in the draw sequence of mixture means. The marginalized sampler exhibits considerably less autocorrelation. The full sampler has autocorrelation function that implies a relative efficiency factor of less than 4 vis-à-via the marginalized sampler. Given that the current implementation of the marginalized sampler is at least 10 times slower than the full sampler, I conclude that the marginalization does not represent a realizable improvement in inference. Figure 1.23 provides the autocorrelation of the density ordinate value evaluated at the point (1,1). As in the case of the first moment, the marginalized or collapsed sampler exhibits less autocorrelation but the differences are minor compared to the higher computational cost of the marginalized sampler.

2
Dirichlet Process Prior and Density Estimation

One of the principal criticisms that can be leveled against the finite mixture of normals approach is that the approach is not non-parametric. A non-parametric approach would require the ability to accommodate a very large number of normal components and that as N approaches infinity the number of components could increase so as to flexibly fit any density. Although I have demonstrated that the finite mixture approach is feasible with a very large number of multivariate components ($K >$ 10), a fixed K procedure cannot make explicit non-parametric claims. Given that it is difficult to develop a sensible method of expanding K with the sample size, consideration of models with potentially an infinite number of components is desirable. The Dirichlet Process (DP) prior is one such approach that explicitly allows for a countably infinite mixture of components. The DP prior can also be thought of as the limit of a finite mixture of normals model.[1] The limiting experiment that yields the DP prior also reveals implications of the DP prior for the number and size of components.

---

[1] Dirichlet Processes can be used to put priors on a mixture of non-normal components or as a part of a general hierarchical approach. However, the DP prior in the mixture of normals model is one of the most useful applications both because of the interpretation of the DP prior as a limit of the finite mixture of normals model, but also because the use of normal components facilitates computation of quantities required for implementation.

2.1 Dirichlet Processes—A Construction

The Dirichlet Process (DP) was introduced by Ferguson (1974).[2] There are a number of different ways to represent a DP. In the applications considered here, the DP is used as part of a mixture model and is often referred to as a Dirichlet Process Mixture. The formal definition of a DP is as a distribution over probability measures defined on some sigma-algebra (collection of subsets) of a space, $\mathcal{X}$, such that the distribution for any finite partition of $\mathcal{X}$ is a Dirichlet distribution. That is, consider any partition, $A_1, \ldots, A_K$, of $\mathcal{X}$, and denote the DP as the distribution $G \sim DP(\alpha, G_0)$, then

$$(G(A_1), \ldots, G(A_K)) \sim \text{Dirichlet}(\alpha G_0(A_1), \ldots, \alpha G_0(A_K)).$$
$$(2.1.1)$$

The two parameters of the DP are a "precision" or tightness parameter, α, and a base measure, G_0. The idea is that the DP is centered over the base measure with precision, α. Larger values of α mean that the DP is more tightly distributed around G_0. For any measurable subset A of $\mathcal{X}$,

$$E[G(A)] = G_0(A)$$
$$Var(G(A)) = \frac{G_0(A)(1 - G_0(A))}{\alpha + 1}. \qquad (2.1.2)$$

Since a DP is a distribution over probability distributions, we can talk meaningfully about a random variable defined by a given draw or realization of the DP. That is, we can define a random variable generated by the DP as a sequence of two conditional distributions.

$$\theta | G \sim G$$
$$G \sim DP(\alpha, G_0) \qquad (2.1.3)$$

---

[2]I am indebted to Professor Y. W. Teh for this construction of the Polya Urn representation.

The DP induces a marginal distribution on the random variable, θ, by integrating over all possible realizations of G.

$$p(\theta) = \int p(\theta|G)p(G)dG \qquad (2.1.4)$$

If we have a collection of θ, then we want to compute the joint distribution of this collection which is generated as the marginal by integrating over the DP. The representation of this joint distribution as a sequence of conditionals forms the basis of the Blackwell-MacQueen Polya Urn model (Blackwell and MacQueen (1973)).

$$p(\theta_1, \ldots, \theta_n) = p(\theta_1)p(\theta_2|\theta_1)\ldots p(\theta_n|\theta_1, \ldots, \theta_{n-1}) \qquad (2.1.5)$$

To obtain all of the component distributions in (2.1.5), we must start by finding the marginal of θ implied by the representation in (2.1.3). This will provide the first term, $p(\theta_1)$. The other terms will be computed from the posterior of a DP. In order to determine the marginal distribution of the first element of θ implied by the DP, we rely on the fact that the Dirichlet distribution is conjugate to the multinomial distribution.

If

$$\pi \sim \text{Dirichlet}(\alpha)$$
$$z|\pi \sim \text{MN}(\pi), \qquad (2.1.6)$$

then

$$z \sim \text{MN}\left(\frac{\alpha_1}{\sum_j \alpha_j}, \ldots, \frac{\alpha_K}{\sum_j \alpha_j}\right)$$
$$\pi|z \sim \text{Dirichlet}(\alpha_1 + \delta_1(z), \ldots, \alpha_K + \delta_K(z)). \qquad (2.1.7)$$

The $\delta_j()$ functions are indicators of whether or not $z = j$. Because of the definition that G defines a Dirichlet distribution over any partition of the sample space, we can use the results of the Dirichlet-Multinomial conjugacy to infer the marginal distribution of θ given in (2.1.4). Given a partition of $\mathcal{X}$, the definition of the DP states that

$$(G(A_1), \ldots, G(A_K)) \sim \text{Dirichlet}(\alpha G_0(A_1), \ldots, \alpha G_0(A_K))$$

$$\Pr(\theta \in A_k | G) = G(A_k).$$

By Dirichlet-Multinomial conjugacy, this implies

$$\Pr(\theta \in A_k) = G_0(A_k)$$

$$(G(A_1), \ldots, G(A_K)) | \theta \sim \text{Dirichlet}(\alpha G_0(A_1)$$

$$+ \delta_\theta(A_1), \ldots, \alpha G_0(A_K) + \delta_\theta(A_K)).$$

If we take a very fine partition, then we can see that the marginal distribution of θ is the base measure for the DP.

$$p(\theta)d\theta = G_0(d\theta)$$

It is also apparent that the posterior distribution of G is also a DP.

$$G|\theta \sim \text{DP}\left(\alpha + 1, \frac{\alpha G_0 + \delta_\theta}{\alpha + 1}\right) \qquad (2.1.8)$$

Here the notation $\frac{\alpha G_0 + \delta_\theta}{\alpha + 1}$ means that the base measure for the posterior DP is a mixture of a draw from the base measure G_0 (with probability $\frac{\alpha}{\alpha + 1}$) and a point mass at θ. To find the distribution of $\theta_2 | \theta_1$, we must integrate out the appropriate posterior DP from the joint distribution.

$$\theta_2 | G^*(\theta_1) \sim G^*(\theta_1)$$

$$G^*(\theta_1) \sim \text{DP}\left(\alpha + 1, \frac{\alpha G_0 + \delta_{\theta_1}}{\alpha + 1}\right)$$

Again, using the Dirichlet-Multinomial results, we can show that the "marginal" or $\theta_2|\theta_1$ has the form

$$\theta_2|\theta_1 \sim \frac{\alpha G_0 + \delta_{\theta_1}}{\alpha + 1}$$

and

$$G|\theta_1, \theta_2 \sim \mathrm{DP}\left(\alpha + 2, \frac{\alpha G_0 + \delta_{\theta_1} + \delta_{\theta_2}}{\alpha + 2}\right).$$

Rolling this idea forward we see that we have obtained the conditional distribution of

$$\theta_n|\theta_1, \ldots, \theta_{n-1} \sim \frac{\alpha G_0 + \sum_{j=1}^{n-1} \delta_{\theta_j}}{\alpha + n - 1}. \tag{2.1.9}$$

This result is often called the Blackwell-MacQueen Polya Urn representation.[3] (2.1.9) provides the practical basis to draw from both the DP prior and posterior and also provides a strong intuition for the sort of joint distribution induced by the DP. Although the marginal of any θ is the base measure, the draws of θ are not independent. In particular, there is a "clustering" or clumping of the thetas which are drawn in the Blackwell-MacQueen representation. That is, it is possible to "repeat" the same θ value.

What is, perhaps, less obvious is that the clustering favors large "clumps" or clusters. That is, if there are I^* unique values drawn in the sequence, $\theta_1, \ldots, \theta_{n-1}$, then those unique values with a higher number of repeats or members of that cluster will

---

[3]The Dirichlet distribution can be obtained from a Polya Urn scheme in which an urn with colored balls is drawn from with a rule that for each ball drawn at random from the urn two balls of the same color are added to replace the ball drawn. The DP can be viewed as a limit of the Polya Urn scheme where the balls take on a continuum of colors.

be drawn with a higher probability when drawing θ_n. Let n_i, $i = 1, \ldots, I^*$ be the number of each unique value of θ drawn. Then the probability that θ_n will be unique value i is proportional to n_i.[4]

2.2 Finite and Infinite Mixture Models

In the previous section, I emphasized the close relationship between the Dirichlet distribution and the DP in providing a construction of the Blackwell-MacQueen Polya Urn representation. Therefore, it should not be surprising that there is a close relationship between the finite mixture model and an interpretation of the DP as an infinite mixture. It is known that draws from the DP are discrete and, therefore, can be used as mixing distributions with countably infinite support. The DP model can be shown as the limit of finite mixture models (see Neal (2000)). It is instructive to follow this limiting argument as it provides some insight as to the way in which the DP works as a prior in a mixture model.

Consider the standard finite mixture model but with a particular parameterization of the Dirichlet prior.

$$
\begin{aligned}
y_i | z_i, \Theta &\sim f\left(y | \theta_{z_i}\right) \\
z_i | \pi &\sim \mathrm{MN}(\pi) \\
\theta &\sim G_0\left(\lambda\right) \\
\pi &\sim \mathrm{Dirichlet}(\alpha/K, \ldots, \alpha/K)
\end{aligned}
\tag{2.2.1}
$$

The notation here is chosen to facilitate the limiting arguments. The base distribution, f, is the multivariate normal distribution. Θ corresponds to the collection of K normal distribution parameters and G_0 is the natural conjugate prior given in (1.4.3) and

---

[4]For this reason, DPs are considered a special case of a more general class of processes called the Chinese Restaurant Process in which clusters grow in proportion to size—which some liken to a person choosing a communal table at a Chinese restaurant. Those tables with larger number of patrons are more likely choices for the nth customer to arrive at the restaurant.

assumed to be independent across the collection of K normal parameters. I explicitly include the hyper-parameters (λ) in the notation for the base distribution, $G_0(\lambda)$, to emphasize that we are using the same prior on the normal component parameters in both the finite and DP mixture approaches.

The Dirichlet distribution is parameterized with base measure, $\frac{1}{K}$, and "concentration" or tightness parameter α. DP model for mixtures is a logical extension of this model.

$$
\begin{aligned}
y_i|\theta_i &\sim f(y|\theta_i) \\
\theta_i|G &\sim G \\
G &\sim \mathrm{DP}(\alpha, G_0(\lambda))
\end{aligned}
\tag{2.2.2}
$$

In order to explore the limit of the Dirichlet finite mixture model as $K \to \infty$, we look at the conditional distribution of $\theta_n|\theta_1, \ldots, \theta_{n-1}$ for both models and show that these conditional distributions are the same in the limit. We begin by reviewing this conditional distribution for the finite mixture model. In the finite mixture model, the conditional distribution is specified via the conditional distribution of the indicator variables. That is, we can think of the prior implied in the finite mixture model as drawing possible values of θ from the base measure or the natural conjugate prior, G_0. There are up to K possible values of θ. However, the Dirichlet prior puts prior mass on models with fewer than K components. This "clustering" of θ values is much like the clustering we have displayed by the marginal process generated by the DP model. To compute the conditional distribution of $\theta_n|\theta_1, \ldots, \theta_{n-1}$, properties of the updating of a Dirichlet distribution given multinomial data are required.

2.2.1 The Predictive Distribution in a Dirichlet-Multinomial Model

The finite mixture model prior can be simulated from by sequentially drawing θ values via an expanding set of conditional draws in the same manner as the DP model. That is, in the finite mixture model, the conditional distribution of the multinomial indicator variables determines the nature of the joint discrete distribution of the θ values, each one of which has a marginal distribution given by the base prior, G_0. In particular, the predictive distribution of the indicators is required.

$$p\,(z_n|z_1,\ldots,z_{n-1}) = \int p\,(z_n|\pi)\,p\,(\pi|z_1,\ldots,z_{n-1},\alpha)\,d\pi$$

$$Pr\,(z_n = j|z_1,\ldots,z_{n-1}) = \int \pi_j\,p\,(\pi|z_1,\ldots,z_{n-1},\alpha)\,d\pi$$

$$(2.2.3)$$

The value in (2.2.3) is simply the mean of the posterior distribution of π which is the mean of a Dirichlet distribution by Dirichlet-Multinomial conjugacy. If

$$\pi \sim \text{Dirichlet}(\alpha, M)\,,$$

then

$$\pi|z \sim \text{Dirichlet}(\alpha^*, M^*)$$
$$\alpha^* = \alpha + n - 1$$

$$M^* = \frac{\alpha M + (n-1)\,\hat{F}}{\alpha + n - 1}$$
$$m_j^* = \frac{\alpha m_j + \sum_{i=1}^{n-1} \delta\,(z_i = j)}{\alpha + n - 1}.$$

$m_j = \frac{1}{K}$. $\hat{F}$ is the empirical distribution of the z vector. Thus,[5]

$$Pr\,(z_n = j|z_1,\ldots,z_{n-1}) = m_j^* = \frac{\frac{\alpha}{K} + \sum_{i=1}^{n-1} \delta\,(z_i = j)}{\alpha + n - 1}. \quad (2.2.4)$$

---

[5]The expected value of jth element of a Dirichlet random vector in the parameterization, $p\,(\pi) \propto \prod_{i=1}^{n-1} \pi_i^{\alpha m_i}$, is simply m_j.

2.2.2 The Limit of the Finite Mixture Model

(2.2.4) shows the conditional probability that the nth indicator takes on the value j given that at least one of the first $n - 1$ indicators takes on the value j.

$$Pr\left(z_n = j | z_1, \ldots, z_{n-1}, z_i = j, \text{ for some } i < n\right)$$
$$= \frac{C_j\left(z_1, \ldots, z_{n-1}\right) + \frac{\alpha}{K}}{\alpha + n - 1}$$
$$Pr\left(z_n = j | z_1, \ldots, z_{n-1}, z_i \neq j, \forall i < n\right)$$
$$= 1 - \sum_{i=1}^{K} \frac{C_i\left(z_1, \ldots, z_{n-1}\right) + \frac{\alpha}{K}}{\alpha + n - 1}$$
$$= 1 - \frac{n - 1 - \frac{\alpha}{k}\left(n - 1\right)}{\alpha + n - 1} = \frac{\alpha - \frac{\alpha}{K}\left(n - 1\right)}{\alpha + n - 1}$$

$C_j\left(z_1, \ldots, z_{n-1}\right)$ counts the number of occurrences of j in the argument vector. If we let $K \to \infty$, then the probability of any one value of the indicator approaches zero and the limit of the conditional probabilities is given by

$$\lim_{K \to \infty} Pr(z_n = j | z_1, \ldots, z_{n-1}, z_i = j, \text{ for some } i < j)$$
$$= \frac{C_j\left(z_1, \ldots, z_{n-1}\right)}{\alpha + n - 1} \tag{2.2.5}$$

$$\lim_{K \to \infty} Pr\left(z_n = j | z_1, \ldots, z_{n-1}, z_i \neq j, \forall i < j\right) = \frac{\alpha}{\alpha + n - 1}. \tag{2.2.6}$$

These limits are the Blackwell-MacQueen conditional probabilities of drawing from either one of the existing values of θ_i or the base measure, G_0 (2.1.9). Thus, the finite mixture model tends to the DP model as the number of components tends to infinity. Both models have (at least the limiting) property that clusters or "clumps" of θ values of large size have higher prior probability of being drawn in the conditional sequence of draws. α determines the probability that a draw will be made from the base measure. In the DP model, the probability of a new value of θ being drawn increases with α. This means that large values of α put higher prior probability on models with more components. If α is small, then both the DP and finite Dirichlet mixture models will put high prior probability on a small number of unique values of θ. It remains to be seen if the finite Dirichlet mixture model rapidly approaches the limiting DP model so that these limiting arguments are useful as a practical matter.

2.3 Stick-Breaking Representation

We have seen that the DP and related finite mixture models have a clustering property that can be seen using the Blackwell-MacQueen Polya Urn representation of a DP model. These clusters are driven by the size of the α parameters. The most direct way of establishing this property of the DP is the so-called "stick-breaking" representation introduced by Sethuraman (1994). The stick-breaking representation is a direct construction of DP draws as countably infinite discrete distributions and proceeds by drawing an infinite sequence of θ_i values from the base measure, G_0. DP draws are then constructed by a discrete distribution on these infinite sequences with weights given by

$$\pi_k = \beta_k \prod_{i=1}^{k-1} (1 - \beta_i) \quad \beta_k \sim \text{Beta}(1, \alpha) . \qquad (2.3.1)$$

Sethuraman shows that a draw from the DP can be constructed from the discrete distribution that puts mass π_k on draws θ_k from G_0.

$$\theta_k \sim G_0$$

$$G = \sum_{k=1}^{\infty} \pi_k \delta_{\theta_k} \qquad (2.3.2)$$

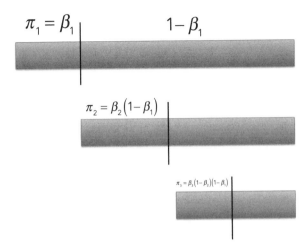

$$\pi_1 = \beta_1 \qquad 1 - \beta_1$$

$$\pi_2 = \beta_2\left(1 - \beta_1\right)$$

$$\pi_3 = \beta_3\left(1 - \beta_2\right)\left(1 - \beta_1\right)$$

Figure 2.1. Stick-Breaking Process

The term "stick-breaking" results from the idea that we start with a stick of length 1, we then draw β_1 from a Beta$(1, \alpha)$ distribution. We break off the left β_1 portion of the stick and then break off the fraction β_2 of the remaining portion, etc. Figure 2.1 depicts the stick-breaking process.

The "sticking-breaking" multiplicative process results in weights which decline geometrically fast in k. Figure 2.2 shows draws from the stick-breaking process for two values of α. For small values of α, the weights decline vary rapidly meaning that the DP generates distributions which put mass on a very small number of unique values of θ. Larger values of α are associated with weights that do not decline as rapidly, allowing for mass to be placed on a larger number of unique values. In sum, the stick-breaking representation shows very clearly how the DP, if used as a prior, creates a dependent joint distribution of θ_i random variables. The dependence induced by the DP prior is created by clustering or clumping the joint distribution of the θ values on a small number of unique values.

alpha = .1

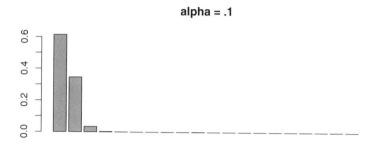

alpha = 5

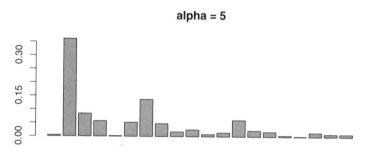

Figure 2.2. Draws of Stick-Breaking Weights

2.4 Polya Urn Representation and Associated Gibbs Sampler

The Polya Urn representation of Blackwell and MacQueen (2.1.9) follows directly from the definition of the DP and can be applied as the basis for defining a Gibbs sampler to draw from the implied prior on the θ parameters or to define a Gibbs sampler for drawing from the posterior of the θ. The Polya Urn representation can be directly used to sample from the DP prior distribution. To sample from the posterior, we must provide the appropriate updating of the Polya Urn scheme conditional on the observed data.

$$\theta_j | Y, \theta_{-j} = \theta_j | y_j, \theta_{j-1} \tag{2.4.1}$$

This simplification of the conditional posterior distribution follows from the conditional independence of the observations given θ.

$$p\left(\theta|Y\right) \propto p\left(Y|\theta\right) p(\theta) = p\left(Y|\theta\right) p\left(\theta_j|\theta_{-j}\right) p\left(\theta_{-j}\right)$$

This implies

$$p\left(\theta_j|\theta_{-j}, Y\right) \propto \prod_{i=1}^{N} p\left(y_i|\theta_i\right) p\left(\theta_j|\theta_{-j}\right) \propto p\left(y_j|\theta_j\right) p\left(\theta_j|\theta_{-j}\right)$$

which establishes that the conditional posterior for θ_j depends only on y_j.

The conditional posterior can readily be obtained by recognizing that the prior admits n "models" for the jth observation. With prior probability, $\frac{1}{\alpha+(n-1)}$, we consider each of the $n-1$ possible other values of θ in θ_{-j} and, with prior probability, $\frac{\alpha}{\alpha+(n-1)}$, we consider a new "model" or a draw from $G_0\left(\lambda\right)$. The conditional posterior simply involves an updating of the prior probabilities by considering the likelihood of each of these "n" models.

$$\theta_j|\theta_{-j}, y_j, \lambda, \alpha \sim \frac{q_0 G_0\left(\lambda\right) + \sum_{i \neq j} q_i \delta_{\theta_i}}{q_0 + \sum_{i \neq j} q_i} \qquad (2.4.2)$$

q_0 and q_i are the posterior probabilities of the n "models."

$$q_0 = p\left(M_0|y_j\right) = \int p\left(y_j|\theta_j\right) p\left(\theta_j|\lambda\right) d\theta_j \times p\left(M_0\right)$$

$$= \int p\left(y_j|\theta_j\right) G_0\left(d\theta_j|\lambda\right) \times \frac{\alpha}{\alpha+(n-1)} \qquad (2.4.3)$$

$$q_i = p\left(M_i|y_j\right) = p\left(y_j|\theta_i\right) \times \frac{1}{\alpha+(n-1)} \qquad (2.4.4)$$

q_0 and q_i can be interpreted as the Bayes Factors corresponding to each of the n possible models. Again, the similarity with the finite mixture model is striking. The draws of the indicator variables (1.5.4) depend on the conditional posterior probabilities or Bayes Factors for each of the possible K components.

2.5 Priors on DP Parameters and Hyper-parameters

We have seen that the DP is a generalization of the finite Dirichlet mixture and, as such, provides a way of extending a finite mixture model to an infinite mixture model. The DP replaces the Dirichlet distribution as part of the prior on the clustering process for mixture parameters. As in the finite mixture process, there are two important aspects of the DP prior: (1) the α parameter which influences the number of clusters or unique values of θ that the DP puts substantial prior mass on; and (2) the base measure, G_0. As we have seen, the DP induces a prior on the mixture parameters. This prior is sometimes termed a Dirichlet Process mixture as it involves integrating out over the draws from the DP as in (2.1.4). Antoniak (1974) provides an important result that shows directly how the α parameter determines the probability distribution of the number of unique values from the DP mixture model.

$$Pr\left(I^* = k\right) = \left\|S_n^{(k)}\right\| \alpha^k \frac{\Gamma\left(\alpha\right)}{\Gamma\left(n + \alpha\right)} \qquad (2.5.1)$$

I^* is the number of unique values of θ in a sequence of n draws from the DP prior. $S_n^{(k)}$ are Stirling numbers of the first kind (see Abramowitz and Stegun (1964), p. 824).[6] Figure 2.3 shows how the distribution of the number of unique θ values varies with α for $N = 500$.

---

[6]Recurrence formulae for the Stirling numbers are not useful when $n > 150$ and the approximation, $S_n^{(k)} \doteq \frac{\Gamma(n)}{\Gamma(k)}\left(\gamma + \ln\left(n\right)\right)^{k-1}$, must be used (here γ is Euler's constant).

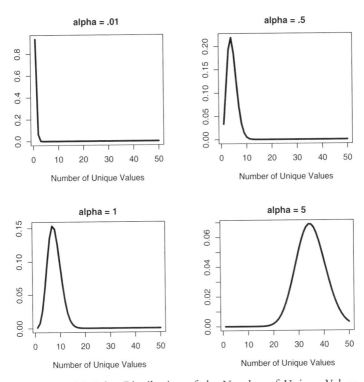

Figure 2.3. DP Prior Distribution of the Number of Unique Values ($N = 500$)

It should be noted that the DP prior will allow for up to N unique values and has full support on 1 to N for any value of α. However, as a practical matter, the support of the distribution is restricted. Small values of α imply a prior which puts almost all mass on only a few unique θ while large values can put the great bulk of the prior mass on 20 or more unique values (see lower right-hand corner of Figure 2.3).

Direct assessment of α may be difficult and the resulting prior may be too tight. For example, with a sample size of 500, I might want to assess a prior that puts substantial mass on from 1 to 20 or some unique values of θ. The unique values of θ correspond to normal components in DP mixture of normals model. If we put a prior on α, we can effectively spread out the prior mass over a larger range of unique components.

$$p\left(I^*\right) = \int p\left(I^*|\alpha\right) p\left(\alpha|\tau\right) d\alpha \qquad (2.5.2)$$

The prior on α must be chosen to facilitate assessment and so that the resulting implied prior on I^* is sensible. In the literature, gamma priors have been used for α (see, for example, Escobar and West (1995)). These priors do provide flexibility but are hard to assess in terms of their implications for the prior distribution of the number of possible components. Conley, Hansen, McCulloch, and Rossi (2008) use a prior of the form

$$p\left(\alpha\right) \propto \left(1 - \frac{\alpha - \underline{\alpha}}{\bar{\alpha} - \underline{\alpha}}\right)^{\phi}. \qquad (2.5.3)$$

Here $\underline{\alpha}$ and $\bar{\alpha}$ are assessed by inspecting how the mode of the distribution of $I^*|\alpha$ varies with α. I^*_{min} and I^*_{max} are assessed by considering the smallest and largest number of unique values we would like to put substantial prior mass on. Intuitively, $\underline{\alpha}$ and $\bar{\alpha}$ are chosen to reflect the range in the number of components that we would like to assess as most probable. For example, if we have a modest data set of only 200 or so observations, we might assess a range of between 1 and 5 components, while if we had much more data, we might assess a larger range. To some Bayesians, this dependence of the prior on the data may seem incoherent. However, in all non-parametric work, we do have the notion that with modest amounts of data, considerable smoothing may be required to produce sensible estimates.

We choose $\underline{\alpha}$ and $\bar{\alpha}$ to correspond to those modes using the approximate Stirling number formula and (2.5.1). [7] It should

---

[7] $\alpha_{mode} = \exp\left(\text{digamma}\left(I_{mode}\right) - ln\left(\gamma + ln\left(n\right)\right)\right)$; digamma () is the derivative of the log-gamma function.

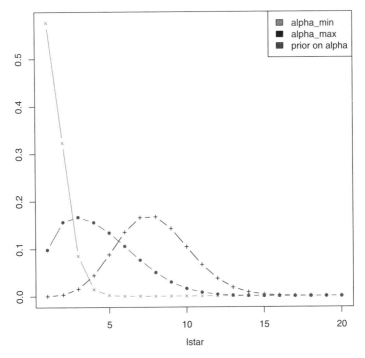

Figure 2.4. Priors on Istar

be emphasized that the prior on I^* created by mixing over the prior in (2.5.3) has support on all values from 1 to the sample size. The power parameter, ϕ, is chosen to spread the prior mass out appropriately.

Figure 2.4 illustrates how I assess the prior for $N = 100$. I assess $I^*_{min} = 1$ and $I^*_{max} = 10$ to provide broad prior support for values from 1 to 10 components. Solving for the implied value of α which sets the mode of the prior distribution of $I^*|\alpha$ to 1 and 10 respectively, I obtain the priors shown by "x" (corresponding to $I^*_{min} = 1$) and by "+" (corresponding to $I^*_{max} = 10$). With $\phi = .5$, the implied prior on I^* obtained from (2.5.2) using prior on α given in (2.5.3) is depicted by •. The mixture prior puts substantial mass on between one and 10 components and does not have the undesirable properties of the priors obtained by fixing α.

The hyper-parameters, λ, of the base distribution, G_0, must also be assessed. Just as the prior over α has an influence over the prior distribution of the number of unique values of θ, it is also true that the hyper-parameters, λ, influence the prior distribution of the number of unique values of θ. λ governs what size and location of normal parameters are proposed. I take the form of G_0 to be the standard natural conjugate prior (1.4.3) used for multivariate normal data. This is the same prior used in the finite mixture of normals and the same considerations in assessing this prior as discussed in section 1.4 apply. λ should be chosen to put prior mass on normal components of varying size (variance) and correlation structure and which cover the support of the data. The role of λ in influencing the posterior distribution of the number of unique values can be seen from the equation for q_0.

$$q_0 = p\left(M_0|y_i\right) = \int p\left(y|\theta\right) p\left(\theta|\lambda\right) d\theta \times \frac{\alpha}{\alpha + (n-1)}$$

This is a standard Bayes Factor for the "model" that θ is a draw from G_0. Recall that here G_0 is represented by the density, $p\left(\theta|\lambda\right)$. The standard intuition for Bayes Factors applies—that is, as the diffusion of G_0 increases, the posterior probability of the model will decline. Therefore, it is not appropriate to assess values of λ that imply a very diffuse G_0. This will reduce the probability that a new value of θ are added to the list of unique values and, effectively, puts an informative prior that favors models with a very small number of components.

Since very diffuse settings for λ are not desirable, some care must be taken if direct assessment of G_0 (λ) is required as in the discussion of the finite mixture model. Another approach would be to place priors on λ and allow the data, in part, to determine the values of λ. This also allows some greater flexibility in the prior. To place priors on λ, G_0 is parameterized as follows:

$$\mu | \Sigma \sim N\left(0, a^{-1}\Sigma\right)$$
$$\Sigma \sim IW(v, vvI_k) \qquad (2.5.4)$$

This parameterization centers on zero for the location and parameterizes the prior on Σ in a way which attempts to isolate tightness and location. As is standard in the IW distribution, the parameter, v, determines the tightness. The location of the IW prior is given by the mode.

$$\text{mode}(\Sigma) = \frac{v}{v+2}vI_d \qquad (2.5.5)$$

This allows the prior to be centered on vI_d where d is the dimension of the data. There is no doubt that this prior parameterization is redundant in that since there are many ways of approximating a distribution there are values of the prior hyperparameters that assess very similar priors. However, the goal is not to conduct precise posterior inference regarding α, v, v, and a, but, instead, to provide adequate flexibility while retaining the benefits of a proper prior full Bayesian approach (diminished over-fitting).

I assess independent priors on each of the parameters, a, v, v.

$$a \sim \text{Unif}(a_l, a^u)$$
$$v \sim \text{Unif}(v_l, v^u)$$
$$v = d - 1 + exp(z)$$
$$z \sim \text{Unif}(v_l, v^u) \qquad (2.5.6)$$

The log-uniform prior on v is assessed to allow for the fact that the implications of changes in the degrees of freedom for an IW distribution differ depending on the size of the degrees of freedom. That is, as degrees of freedom parameter increases the IW distribution tightens at a slower rate. The support of the prior on v is chosen to keep the IW density proper (this requires that $v_l > 0$).

2.6 Gibbs Sampler for DP Models and Density Estimation

A Gibbs sampler for the DP model and our prior parameterization is given by

$$\{\theta_i\} \,|\, Y, a, v, v, \alpha \qquad\qquad (2.6.1)$$

$$\alpha | I^* \qquad\qquad (2.6.2)$$

$$a | \left\{\theta_i^*\right\} \qquad\qquad (2.6.3)$$

$$v | \left\{\theta_i^*\right\}, v \qquad\qquad (2.6.4)$$

$$v | \left\{\theta_i^*\right\}, v. \qquad\qquad (2.6.5)$$

The first conditional in the Gibbs sampler (2.6.1) is the standard posterior Polya Urn representation. The DP hyper-parameters depend (a posteriori) only on the set of unique θ values which I denote as $\left\{\theta_i^*\right\}$. Given the set of unique theta values, the α and the λ are, a posteriori, independent. The conditional posterior of α depends only on the number of unique values (2.6.2), denoted here by I^*. The conditional posterior of the G_0 hyper-parameters, a, v, v, factors into two parts as a is independent of v, v given $\left\{\theta_i^*\right\}$. The form of this conditional posterior is

$$p\left(a, v, v | \left\{\theta_i^*\right\}\right) \propto \prod_{j=1}^{I^*} \phi\left(\mu_j^* | 0, a^{-1}\Sigma_j^*\right) \times$$
$$\text{IW}\left(\Sigma_j^* | v, V = vvI_d\right) p\left(a, v, v\right). \quad (2.6.6)$$

Here $\phi(x|\mu, \Sigma)$ is the multivariate normal density evaluated at point x given mean μ and covariance matrix Σ and $\text{IW}(A|v, V)$ is the Inverted Wishart distribution evaluated at matrix A with parameters, v, V.

As noted in Chapter 1, the Bayesian analogue of density estimation is to compute the predictive distribution of a "future" observation.

$$p(y_{n+1}|Y) = \int p(y_{n+1}|\theta_{n+1}) p(\theta_{n+1}|Y) d\theta_{n=1} \quad (2.6.7)$$

$$p(\theta_{n+1}|Y) = \int p(\theta_{n+1}|\theta_1, \ldots, \theta_n) p(\theta_1, \ldots, \theta_n|Y) d\theta_1 \cdots d\theta_n$$
$$(2.6.8)$$

The integral in (2.6.8) can be computed via simulation by drawing using the Polya Urn representation for $\theta_{n+1}|\theta_1, \ldots, \theta_n$ to draw θ_{n+1} for each posterior draw of $\theta_1, \ldots, \theta_n$. For each of these draws of θ_{n+1}, construct the normal density and average these draws. The algorithm for computing the posterior mean of the predictive density is given by

1. draw $\theta_{n+1}^r|\Theta^r, \lambda^r$;
2. construct $\phi\left(y_{n+1}|\theta_{n+1}^r\right)$;
3. average the normal densities in step 2,
   $\hat{p}(y_{n+1}) = \frac{1}{R} \sum_{r=1}^{R} \phi\left(y_{n+1}|\theta_{n+1}^r\right)$.

Θ in the above steps refers to the collection of $\theta_1, \ldots, \theta_n$. λ is the triple a, v, v. The draws in step 1 use the Polya Urn scheme

$$\theta_{n+1}|\Theta \sim \begin{cases} G_0(\lambda) \text{ with prob } \dfrac{\alpha}{\alpha + n}, \\[2ex] \delta_{\theta_i} \text{ with prob } \dfrac{1}{\alpha + n}. \end{cases}$$

2.7 Scaling the Data

In both the finite mixture and infinite mixture of normals (DP) approaches, the scaling of the data is important for prior assessment. I have explained that priors reflect a priori views regarding the number and "size" of the normal components. Scaling the data facilitates assessment of the key prior hyper-parameters as discussed in sections 1.4 and 2.5. If the bulk of the data can be constrained to a hyper-cube, then the appropriate priors on the location and variance of normal components can be assessed. I have demonstrated that choices of very diffuse priors can be informative in unintended ways and that some care is required in prior assessment so as to retain maximum flexibility in the mixture of normals approach. Scaling of the data provides one way of providing for reasonable default prior settings. I will consider scaling by centering and normalizing the data.

$$Z = (Y - \iota \bar{y}) D \qquad (2.7.1)$$

$\bar{y}$ centers the data and can be taken to be the vector of column means of the data matrix, Y. D is a diagonal scaling matrix which can be formed using the sample standard deviations of the columns of Y. $D = t \begin{bmatrix} 1/s_1 & & \\ & \ddots & \\ & & 1/s_d \end{bmatrix}$. Some might prefer to use the sample ranges and sample medians as the basis for scaling instead. I should emphasize that only an approximate rescaling is required essentially to get a rough centering on zero and a scale of approximately one. Given a mixture of normal density estimate for the scaled data, Z, we can transform back to the original data scale as follows:

$$\mu_k^y = D^{-1} \mu_k^z + \bar{y}$$

$$\Sigma_k^y = D^{-1} \Sigma_k^z D^{-1}$$

2.8 Density Estimation Examples

To illustrate the properties of the infinite mixture or DP density inference procedures outlined above, I will return to two examples used to illustrate the Finite Mixture (FM) model: (1) a Chi-squared distribution with six degrees of freedom; and (2) a five-dimensional log-normal distribution (1.7.4). Both models do not nest a finite mixture of normals. While there is a limiting relationship between the FM and DP models (see (2.2.2) and (2.2.5)), in practical applications no more than 20 or so components are used in the FM model. In addition, priors have been placed on the hyper-parameters of the G_0 distribution in the DP approach taken here (2.5.6). Thus, there may be performance differences between the FM and DP approaches that are of material importance.

I will begin with the χ_6^2 example. Figure 2.5 shows DP density estimates for a sample of 100 observations. The dark line denotes the true Chi-squared density and the lighter line shows the DP posterior mean estimate of the density using the default prior settings in the bayesm procedure. The default prior settings for the priors on the G_0 are given by

$$a_l = .01, \ a^u = 10, \ v_l = .1, \ v^u = 4 \ v_l = .01, \ v^u = 3. \quad (2.8.1)$$

The upper panel of Figure 2.5 shows the density estimate for the default prior settings. To the right of the density estimate is the distribution of I^*, the number of unique components drawn. In the middle panel of the figure, I show the density estimate with a very informative prior set so as to force the DP prior to put high prior probability on models with a large number of "small" (e.g., low variance) components. These "small component" settings are given by

$$a_l = .01, \ a^u = 2, \ v_l = .1, \ v^u = .1 \ v_l$$
$$= dim\,(y) - 1 + 50, \ v^u = dim\,(y) - 1 + 50. \quad (2.8.2)$$

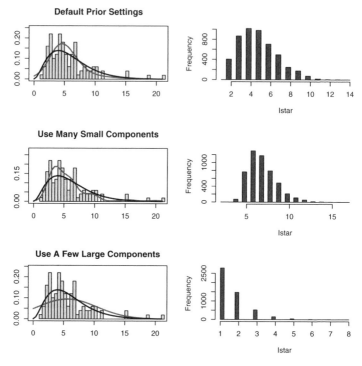

Figure 2.5. Chi-squared Example

This prior is dogmatic regarding the IW degrees of freedom and V matrix parameters, allowing only components with "small" values of the normal component variance-covariance matrix to be proposed. The prior on the DP tightness parameter, α, has not been changed. The middle panel shows a much more "lumpy" density estimate similar to what is obtained by many kernel smoothing procedures. There is also a much larger number of components "active" in the DP procedure. This is due to the prior which favors small components. To approximate the skewed Chi-squared distribution, more "small" components are required in the default setting case where the prior put mass on components in a wider range of sizes. In the bottom panel of the figure, prior settings which put mass mostly on large components are considered. The "large" component settings are

$$a_l = .01,\ a^u = 2,\ v_l = 4,\ v^u = 4\,v_l$$
$$= dim\,(y) - 1 + 50,\ v^u = dim\,(y) - 1 + 50.$$
$$(2.8.3)$$

This dogmatic prior hampers the flexibility of the DP prior representation by assigning low prior probability to small and medium sized normal components. As discussed in section 2.7, the data is scaled so that there is some meaning to the notion of "small" and "large" components as scaling means that most data points lie in the unit cube defined by the interval, $(-3, 3)$. The prior settings in (2.8.3) result in a much more "normal" and less skewed density estimate. The DP posterior uses a far smaller number of components—essentially "giving up" on more than a few normal components because the prior limits proposals to only relative "large" components. That is because the only way that mixtures of normals can accommodate skewness is by "assigning" components with somewhat smaller variance to put more mass in the right tail. In this example, the prior inhibits this aspect of the DP approximation. It should be emphasized that the prior settings (2.8.2) and (2.8.3) are very extreme priors chosen only to illustrate the role of the prior hyper-parameters. No reasonable researcher would impose such severe prior views.

While it is possible to use extreme settings for the IW prior parameters that very much limit or hamper the approximation properties of the DP prior as used here, the DP density estimates are much less influenced by priors on the α and a parameters. Recall that α is the DP prior tightness parameter that governs the prior probability of various numbers of components. My approach is to use a prior on α (2.5.3) which is not dogmatic in the implied prior distribution of the number of normal components. The hyper-parameters of this prior are chosen on the basis of the minimum and maximum number of components

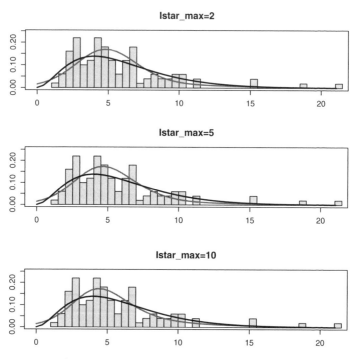

Figure 2.6. Sensitivity to the I^*_{max} Prior Parameter

expected. It should be emphasized, however, that the prior does not impose limits on the posterior support of I^*, rather the prior simply makes some values of I^* less probable. Figure 2.6 shows the same data generated from the χ^2_6 distribution but with small and large values of the I^*_{max} prior hyper-parameter. The DP density estimate does not change over a wide range of this prior parameter.

The DP density estimates are also not strongly influenced by the value of the a parameter. Figure 2.7 shows the sensitivity of the Chi-squared example density estimates to the precision parameter in the G_0 prior. Here I have adopted the standard default prior settings for all other hyper-parameters and set the

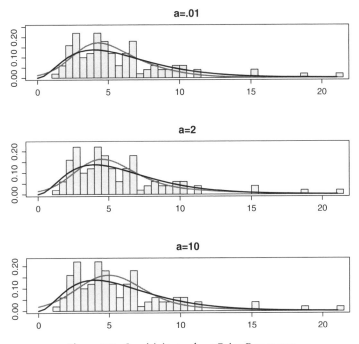

Figure 2.7. Sensitivity to the *a* Prior Parameter

a precision parameter to a range of values from very small to very large. The DP density estimates do not change materially as the *a* parameter is varied. Given that I put priors on the other G_0 parameters, v and ν, this is not surprising. If *a* is set to a large value, then the prior mass on the μ normal component parameters depends on the range of values from a mixture of IW distributions. As long as the priors for v and ν are sufficiently diffuse, the reduced flexibility from a high value of *a* will be limited. Even with a large value of *a*, the priors used here will put mass on a wide range of possible Σ values.

While the Chi-squared examples provide some reassurance that the DP prior hyper-parameters settings are reasonable, the true test is an example of multivariate density estimation.

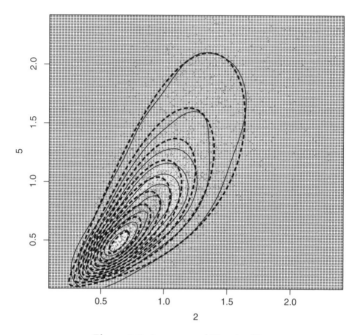

Figure 2.8. Log-normal Density Fit

I return to the five-dimensional log-normal distribution considered in section 1.7. Figure 2.8 shows bivariate marginal on two
of the five dimensions implied by the DP posterior mean density
estimate (black) and the true bivariate marginal (dashed). The
DP procedure does a good job of approximating highly skewed
and non-normal log-normal distribution. The only area where
the DP-based estimate differs from the true density is toward the
origin where there is little data.

Figure 2.9 compares the DP-based density estimate to that
obtained from a 20 component FM model with default prior
parameters. The default prior parameters are

$$a^* = 1/\kappa; \quad a = .01; \quad \nu = 8; \quad V = \nu I. \qquad (2.8.4)$$

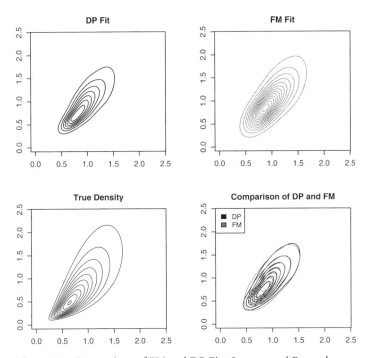

Figure 2.9. Comparison of FM and DP Fits: Log-normal Example

The DP fit shows greater skewness and better approximation to the contours of the true bi-variate marginal density. This might appear to contradict the limiting result that FM models converge to the DP prior. One possibility is that $K = 20$ is not near the limiting case. The other is that even though the base measure, G_0, is of the same form for the FM and DP case, I have put priors on the G_0 parameters in the DP case. This means that we have an effectively different and more flexible prior in the DP case. To examine this, I fixed the G_0 parameters at the same defaults as in the FM case. Figure 2.10 shows this comparison. The DP fit now closely resembles the FM fit. As a further test, I fixed the G_0 parameters in the FM prior to the values which are at the posterior modes obtained from the full

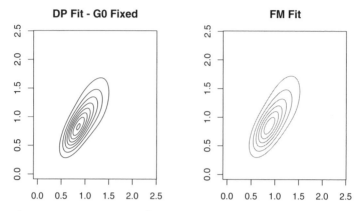

Figure 2.10. Comparison of FM and DP Fits with G_0 Parameters Fixed

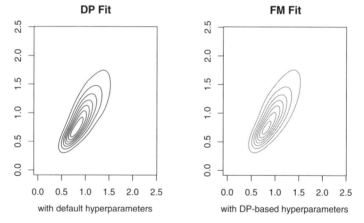

Figure 2.11. Comparison of FM and DP Fits: G_0 Parameters Set at DP Posterior Mode

DP posterior which allows for posterior inference regarding the G_0 parameters. Figure 2.11 shows that the G_0 parameter settings do have a material influence on the FM approximation for multivariate examples which are very non-normal. The posterior on the G_0 parameters derived from the DP procedure finds values of a at about .75 instead of the default of .01. More importantly, the default value of $V = 8I$ used in the IW prior in the FM procedure is now about $V = .9I$ based on the DP posterior. This means that the default prior settings for the FM procedure are set to put too little prior mass on models with "small" normal components. For this reason, I recommend that priors be placed on the G_0 hyper-parameters for the purpose of imparting sufficient flexibility to the prior. Default "diffuse" settings for the prior parameters in these models are, in fact, highly informative in the sense that they can put very high prior probability on models with large normal components, limiting the flexibility of either a finite or infinite mixture approach.

3

Non-parametric Regression

3.1 Joint vs. Conditional Density Approaches

From a non-parametric point of view, a regression model is simply a model for the conditional distribution of the dependent variable, y, given the vector of independent variables, x. A full non-parametric approach should directly estimate or make inferences regarding this conditional distribution and, therefore, provide estimates of any functional of this distribution. Much of the large literature on non-parametric regression models focuses on the conditional mean function.

$$\mathbb{E}[y|x] = \int y p\,(y|x)\,dy \qquad (3.1.1)$$

Various methods have been proposed to make the specification of the conditional mean function as flexible as possible and verify various non-parametric claims that as the sample size increases without bound the approximation method proposed produces a consistent estimate of the true conditional mean function. In this literature, inference regarding the conditional mean function requires an artful application of asymptotic methods. Evaluating the asymptotic approximation is not always easy and a bootstrap is often advocated as a way of conducting inference.

The conditional mean is only one possible summary of the conditional distribution. For non-normal conditional distributions, it is not clear that the conditional mean is the most relevant quantity. The conditional quantiles may be more useful not only as a summary of the central tendency of the conditional distribution but also of the conditional behavior of the tails of the distribution of y. $q_p(x)$ denotes the pth conditional quantile and is the solution to an implicit equation.

$$p = \int_{-\infty}^{q_p(x)} p(y|x)\, dy \qquad (3.1.2)$$

A substantial literature on quantile regression has been developed. In much of this literature, an estimator is proposed for a specified quantile and there is no natural way to impose the restriction that estimated conditional quantiles should be increasing in p given x. In particular, the requirement that an estimation procedure provide a consistent estimate of the conditional quantile function for a given quantile (choice of p) does not require that conditional quantile estimates be logically consistent in the sense that they shouldn't cross for any given fixed sample size.

What I term a fully non-parametric approach to regression takes as the object of interest for inference the entire conditional distribution of y given x. There are two approaches in the Bayesian literature which I would term as "fully non-parametric" and utilize mixture of normals models. The first approach (exemplified by Geweke and Keane (2007) and Villani, Kohn, and Giordani (2009)) models the conditional distribution. The second approach (Taddy and Kottas (2010)) models the joint distribution of y and x and then computes the conditional distribution of $y|x$ implicit in the fitted joint. Either the first or "conditional" approach or the second or "joint" approach can be implemented with either finite or infinite mixtures of normals. The appeal of the conditional approach stems from the fact that it appears to affect a dimension reduction in that the conditional distribution is only univariate. However, the conditional approach must specify a way in which the conditioning variables affect the distribution of y and this must be done in a sufficiently flexible way that a non-parametric claim can be substantiated. In particular, the conditional approach must allow for a very general non-linear dependence of the conditional mean on x and a very general form of conditional heterogeneity such as, but not limited to, conditional heteroskedasticity. Geweke and Keane (2007) allow for the mixture probabilities in a finite mixture of normals to depend on the x variables. This allows for conditional heterogeneity as well as non-linearity of the conditional mean function. The essence of the Geweke and Keane (2007) contribution is a model of the form

$$y|x \sim \sum_{k=1}^{K} \pi_k(x)\, \phi\left(x'\beta_k, \sigma_k\right). \tag{3.1.3}$$

Villani, Kohn, and Giordani (2009) extend the approach of Geweke and Keane (2007) to include a scaling function that creates conditional heteroskedasticity directly instead of relying on the influence of x on the mixture probabilities.[1]

$$y|x \sim \sum_{k=1}^{K} \pi_k(x)\, \phi(x'\beta_k, \sigma_k \exp(\delta_k'x)) \tag{3.1.4}$$

For full flexibility, the conditional approach must allow for the x variables to affect at least the mixing probabilities. The conditional approach can easily destroy the conditional conjugacy that enables a full Gibbs sampling approach in the case of the standard mixture of normals model (both authors advocate a Hybrid method which include Metropolis steps which require tuning). While the Gibbs sampler for either the finite or infinite mixture of normals models works well even in high dimensions, Metropolis methods can break down in high dimensions. In addition, Villani, Kohn, and Giordani (2009) point out that, even though they have what they term a "very efficient" MCMC method, data sets of more than 20,000 or so observations are "onerous." This poses a real problem as, by their very nature,

---

[1]In addition, they allow for possibly different subsets of the x variables to affect the mixture probabilities, the means of the mixture components, and the scaling function. In effect, they implement variable selection for which a set of x variables affect each part of the mixture of normals model.

non-parametric methods are most applicable to very large data sets. For the normal mixture approach advocated here, data sets in excess of 100,000 observations can be analyzed in less than one-half hour of computing on pedestrian equipment.

In contrast, the joint approach does not require assumptions and specific functional forms for how the x variables influence the conditional distribution of y. For the joint approach a full Gibbs sampler is available for both the finite and infinite mixture models. All forms of conditional heterogeneity and non-linearity are automatically accommodated in the joint distribution. In addition, the simply additive forms for the mixture of normal densities facilitate computation of conditional means and quantiles implied by the joint distribution. This provides a very practical approach to non-parametric regression. Although full non-parametric claims have not been verified for the conditional approach, the existing literature on Bayesian density estimation for DP prior models as well as the limiting relationship between finite mixture and DP prior models establishes that the joint approach will consistently recover the joint density (and, therefore, the conditional associated with the joint). Finally, MCMC methods for standard finite or infinite mixture of normals models are fast enough to analyze very large data sets.

3.2 Implementing the Joint Approach with Mixtures of Normals

The joint approach begins with a mixture of normals approximation to the joint distribution of (y, x). Either the finite or infinite mixture approach facilitates a Gibbs sampler which provides draws from the posterior distribution of the joint distribution.

$$f(y, x)^r = \sum_{k=1}^{K} \pi_k^r \phi\left(y, x | \mu_k^r, \Sigma_k^r\right) \qquad (3.2.1)$$

Each draw in (3.2.1) can be used to compute the implied conditional distribution.

$$f(y|x)^r = \frac{\sum_{k=1}^{K} \pi_k^r \phi\left(y, x | \mu_k^r, \Sigma_k^r\right)}{f(x)^r} \tag{3.2.2}$$

$$f(x)^r = \int f(y, x)^r \, dy = \sum_{k=1}^{K} \pi_k^r \bar{\phi}_k(x)^r \tag{3.2.3}$$

$$\bar{\phi}_k(x)^r = \int \phi\left(y, x | \mu_k^r, \Sigma_k^r\right) dy \tag{3.2.4}$$

Any functional of that conditional distribution such as the conditional mean and conditional quantiles can be computed based on the rth draw of the joint. In this manner, the posterior distribution for any desired functional involving the conditional distribution can be obtained. The linear structure of the mixture of normals model can be exploited to facilitate computation of the conditional mean and quantiles. Below I will suppress the r draw superscript.

$$\mathbb{E}[y|x] = \int y f(y|x) \, dy = \int y \frac{\sum_k \pi_k \phi_k(y, x)}{f(x)} dy$$

$$= \frac{1}{f(x)} \int y \sum_k \pi_k \phi_k(y, x) \, dy$$

$$= \frac{1}{f(x)} \sum_{k=1}^{K} \pi_k \int y \phi_k(y, x) \, dy$$

$$= \frac{1}{f(x)} \sum_{k=1}^{K} \pi_k \int y \frac{\phi_k(y, x)}{\bar{\phi}_k(x)} \bar{\phi}_k(x) \, dy$$

$$= \frac{1}{f(x)} \sum_{k=1}^{K} \pi_k \mathbb{E}_k[y|x] \tag{3.2.5}$$

Here I use the notation, ϕ_k, to refer to the kth normal component (suppressing the mean and covariance parameters). $\mathbb{E}_k[\ ]$ is the conditional mean for the kth normal component.

$$\mathbb{E}_k[y|x] = \mu_{y,k} - F_k'\mu_{x,k} + F_k'x \qquad (3.2.6)$$

The elements needed to compute the conditional mean in (3.2.6) can be read off the appropriate elements of the inverse of the covariance matrix, Σ_k (a d $= 1 + \dim(x)$ dimensional matrix).

$$F_k' = -\frac{1}{\sigma_k^{1,1}}\gamma_{1,k}[2:d] \qquad (3.2.7)$$

$$\Sigma_k^{-1} = \begin{bmatrix} \gamma_{1,k}' \\ \vdots \\ \gamma_{d,k}' \end{bmatrix} \qquad (3.2.8)$$

Here I assume that the first element of the mixture of normals approximation to the joint distribution of (y, x) corresponds to y and the other $dim(x)$ elements correspond to the x variables. $\sigma^{1,1}$ is the $(1, 1)$ element of Σ_k^{-1}. The γ vectors are the rows of Σ_k^{-1}.

In a non-parametric setting, the derivative of the conditional mean function is also useful as a measure of the local effect size. Again, the mixture of normals form facilitates a simple formula for the gradient of the conditional mean function.

$$\frac{\partial \mathbb{E}[y|x]}{\partial x} = \frac{1}{f(x)}\sum_{k=1}^{K}\pi_k\bar{\phi}_k(x)\left[F_k' - (\mathbb{E}_k[y|x]\right.$$
$$\left. -\mathbb{E}[y|x])\Sigma_{xx,k}^{-1}(x - \mu_{x,k})\right] \qquad (3.2.9)$$

Here $\Sigma_{xx,k}$ is the submatrix of Σ_k corresponding to the x variables.

Conditional quantiles can be computed as well by inverting the CDF of the conditional distribution (3.1.2) using a search algorithm. The mixture of normal form facilitates the computation of the CDF of the conditional distribution of $y|x$.

$$p = \int_{-\infty}^{q_p} f(y|x)\,dy$$

$$= \frac{1}{f(x)} \int_{-\infty}^{q_p} \sum_{k=1}^{K} \pi_k \phi_k(y, x)\,dy$$

$$= \frac{1}{f(x)} \sum_{k=1}^{K} \pi_k \phi_k(x) \int_{-\infty}^{q_p} \phi_k(y|x)\,dy$$

$$= \frac{1}{f(x)} \sum_{k=1}^{K} \pi_k \phi_k(x) \Phi_{y|x}^{k}(q_p) \tag{3.2.10}$$

$\Phi_{y|x}^{k}$ is the CDF of the conditional distribution of $y|x$ for the kth normal component.

3.3 Examples of Non-parametric Regression Using Joint Approach

To illustrate the functioning of the joint approach, I will start with a simple example that exhibits both a non-linear conditional mean as well as conditional heterogeneity.

$$y = \frac{1}{20}x^3 \varepsilon \tag{3.3.1}$$

$$\varepsilon \sim \chi_{10}^2$$

To simulate data from this model, x is drawn from a uniform distribution on $(1, 4)$ and $N = 500$. Figure 3.1 shows a scatter plot of the data with the posterior mean of the joint density of (y, x) superimposed. The pronounced non-linearity and conditional heteroskedasticity is obvious in the figure. The fit of the joint density of y, x is simply a means to the end of computing various

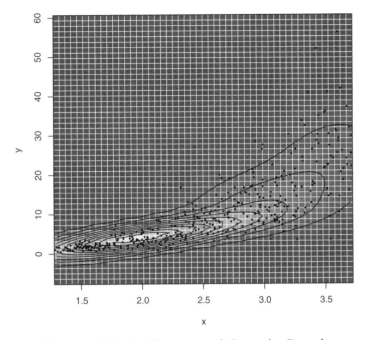

Figure 3.1. Univariate Non-parametric Regression Example

functionals of the conditional distribution of $y|x$. Figure 3.2 shows the true conditional mean as well as the posterior mean of the conditional mean based on the average of the draws of the conditional mean function which is computed for every draw of the joint distribution using 3.2.5. This data set is highly informative regarding the conditional distribution and the conditional mean and there is a close correspondence between true and fitted conditional mean. It is also informative to note how smooth the conditional mean function is. Other approaches such as kernel-based regression and Bayesian approaches such as that advocated in section 10.2 of Koop (2003) require explicit assessment of prior smoothing parameters. Here prior assessment is based on the scaling of the data and our approach which avoids

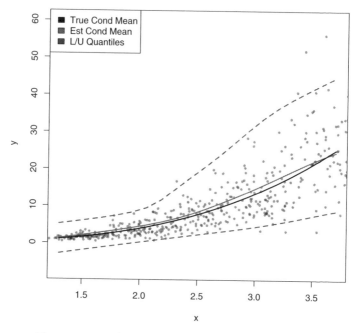

Figure 3.2. Conditional Mean and Conditional Quantiles

absurd priors that favor models with a very small number of components or impose a prior belief that components are very small. The results shown here are based on a 10 component fit but are indistinguishable from a 20 component fit. This lack of over-fitting is a consequence of the use of proper priors for the mixture model as demonstrated in Chapter 1. Figure 3.2 displays the conditional quantiles functions corresponding to .05 and .95. The quantiles functions display how effectively the joint approach captures the conditional heteroskedasticity in this model.

Some would argue that the derivative of the conditional mean function is an important summary of the local affect of x on the conditional mean of y. The true and estimated derivative functions are shown in Figure 3.3. In the range of the data, even

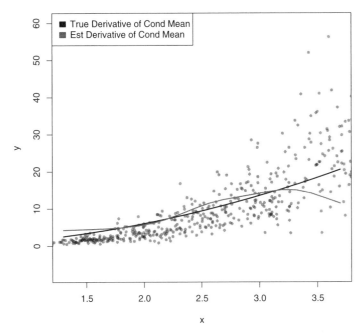

Figure 3.3. Conditional Mean and Derivative

derivatives are reasonably well-estimated which is a somewhat more rigorous test of the procedure as the derivatives are more sensitive to minor deviations of the estimated conditional mean from the true conditional mean. The derivative estimates also illustrate the high degree of smoothness that the procedure imparts to the conditional mean estimate.

The example considered in Figure 3.1 is a very favorable data configuration because of the uniform distribution of the x values. While Bayesian methods can deliver smooth estimates of the fitted conditional mean function even for small data sets, any truly non-parametric method will be sensitive to sparseness in the data. If, for example, the x values are clustered at two points on the axis with a sparse patch in between, then any non-parametric procedure, no matter how powerful, will have

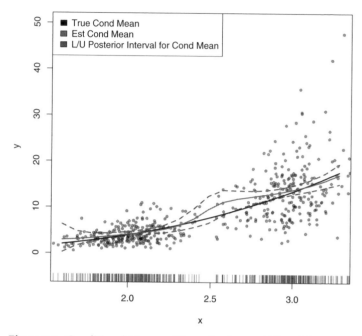

Figure 3.4. Conditional Mean and Posterior Interval: Bimodal *x*

difficulties making precise inferences regarding the conditional distribution in the sparse region. However, one unique advantage of the Bayesian approach is that we obtain at no extra computational costs information about the precision of the posterior inference.

Figure 3.4 shows the same regression model in (3.3.1) but with a bimodal distribution of *x* values. The distribution of *x* values is indicated by the hash marks along the bottom of the figure. This distribution has a "hole" or sparse area around *x* = 2.5. The light-colored dashed lines in the figure correspond to the .025 and .0975 quantiles of the posterior distribution of the conditional mean. The posterior intervals, displayed by the lower and upper dashed lines in Figure 3.4, widen around the "hole" at *x* = 2.5. This reflects appropriately that there is little sample information regarding the conditional mean at this point in the x space. In addition, the posterior intervals widen slightly at either end of the range of x, appropriately reflecting the lack of data beyond the range of x.

The previous example is encouraging in terms of the ease by which the joint approach approximates heterogeneous and non-linear regression relationships. However, the case of more than one independent variable will demonstrate the practical value of the joint approach. The next example is inspired by the aggregate share models considered by Berry (1994) in the sense that the regression error term enters the linear part of logistic style transformation which generates continuous but bounded y data.

$$y = \frac{\exp(1.5x_1 + 1.5x_2 - 1.5x_1x_2 + \varepsilon)}{1 + \exp(1.5x_1 + 1.5x_2 - 1.5x_1x_2 + \varepsilon)} \qquad (3.3.2)$$
$$\varepsilon \sim N(0, .5)$$

In this example, the conditional mean function is not available in closed form but will have non-linearities induced by both the logistic transformation and the interaction term in the model. The error terms are incorporated in a highly non-linear way and the conditional distribution of $y|x_1, x_2$ will display conditional heteroskedasticity. The conditional mean and conditional quantile functions are now surfaces. Data was simulated with $N = 500$ and x_1, x_2 distributed as independent and uniform on $(-1, 1)$. Figure 3.5 shows the conditional mean surface (middle plane) and the 5 and 95 percent quantile surfaces with grid hatching (lower and upper planes, respectively). The spheres represent the data points. The pronounced non-linearity of the surfaces is clear. Figure 3.6 shows the same conditional mean and quantile surfaces from the viewpoint to the side and below the viewpoint shown in Figure 3.5. Both figures illustrate the exceptional smoothing of the joint approach as well as the ability to generate non-crossing quantile surfaces. Figure 3.7 shows the posterior mean of the conditional mean surface along with the posterior interval surfaces corresponding to the .05 and

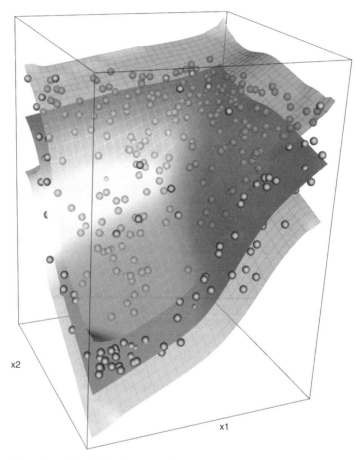

Figure 3.5. Conditional Mean and Quantiles: Bivariate *x* Example

.95 quantiles. The upper and lower planes indicate the small amount of uncertainty in the conditional mean function and show the property that uncertainty increases at the edge of the range of the data in the *x* space.

These two examples illustrate that the joint approach provides a practical non-parametric approach to regression that does not

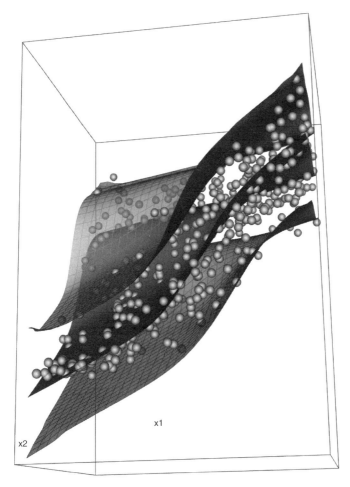

Figure 3.6. Conditional Mean and Quantiles in Bivariate x Example:
Alternative Viewpoint

require specification of a flexible regression function or the nature of the conditional heterogeneity that might be encountered
in the data. Smoothing is achieved via standard priors for the

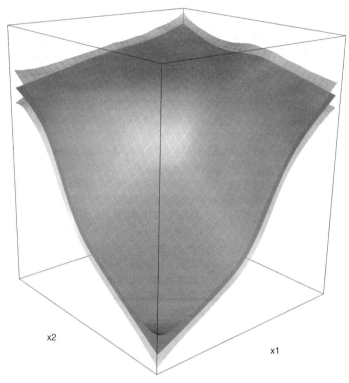

Figure 3.7. Conditional Mean and Posterior Interval: Bivariate *x* Example

mixture of normals model and does not require special tuning for specific data sets.

3.4 Discrete Dependent Variables

The joint approach to conditional distribution inference can easily be extended to the case in which the dependent variable is multinomial using the same mixture of normals tool. The essential identity which facilitates the mixture of normal approach is simply to express the joint of y, x as the weighted average of the conditional distributions of $x|y$.

$$p(y, x) = \Pr(y = j) \, p\,(x|y = j) \ for \ j = 1, \ldots, J \qquad (3.4.1)$$

The conditional distribution is characterized by conditional choice probabilities which are given by

$$Pr\,(y = j|x) = \frac{\Pr(y = j)\, p\,(x|y = j)}{\sum_{j=1}^{J} \Pr(y = j)\, p\,(x|y = j)}. \qquad (3.4.2)$$

Abe (1995) used this identity to develop a non-Bayesian approach to non-parametric choices. Abe used kernel smoothing methods to approximate the conditional densities of $x|y$. Kernel smoothing methods are not practical except in very small (two or fewer) dimensions and great care has to be taken to "tune" the smoothing parameters. In addition, computing reliable sampling errors for estimation of the conditional density requires additional computations and is of dubious quality. (3.4.2) shows the relationship between the conditional choice probabilities and what amounts to a classification or discriminant approach. Consider the problem of classifying a new observation for which we only know the values of x. Classification of that observation is based on our inferences about the distribution of x conditional on y in the complete data set. That is, the distribution of the independent variables conditional on the choice alternative is what enables the determination of the probability that y is a particular choice alternative. In addition, the "base" rate or marginal probabilities of y are important as is standard in Bayes theorem.

If a mixture of normals is used to approximate each of the J conditional densities of $x|y = j$ and the parameters of each of these mixture distributions are a priori independent, then the posterior analysis breaks down into J separate problems with independent Markov chains. The posterior for the marginal distribution of y can also be handled separately if an

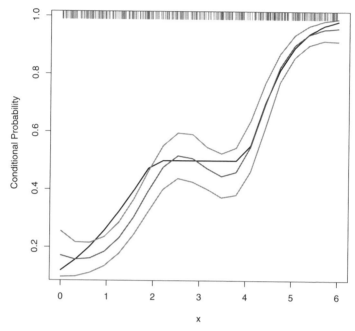

Figure 3.8. Non-parametric Choice Model: N = 500, Uniform *x*

independent prior is used. It will be convenient to use an independent Dirichlet prior for the marginal distribution of *y*. The draws of each of these *J* mixture of normal parameters as well as the probabilities in the marginal distribution of *y* can be used to construct the posterior distribution of the conditional choice probabilities.

I will illustrate the functioning of this approach using a binary choice model which has a flat in the conditional choice probabilities—a feature that is not present in most parametric choice model specifications. I will simulate a sample of 500 from this model with $x \sim \text{Unif}(0, 6)$. Figure 3.8 shows the true conditional mean (probability locus) function as a dark line. The lighter line is the posterior mean of the conditional choice probability and the lightest lines represent the .05 and .95 boundaries of the posterior interval for the posterior mean. The hatch marks on the top margin of the plot indicate the marginal distribution of x. Clearly, the non-parametric method provides a very precise estimate of the conditional choice probability without any need to specify a set of basis functions to approximate how the x value is mapped into an index which is used to compute the conditional choice probability. That is, an alternative approach would to be specify a flexible set of CDF functions which map the index into a probability and then use splines or other methods to accommodate maximum flexibility in the manner in which x drives the index. None of these steps are necessary here.

As in all truly non-parametric methods, the performance of the method demonstrated here depends on the distribution of the x values. If the x distribution is sparse or has limited range, then inference regarding the conditional choice probabilities will be imprecise. The posterior intervals will widen to essentially the entire possible range of probabilities as the x value is moved way from the support of the x distribution. The informativeness of any given size sample is largely driven by the x distribution rather than the number of observations per se. Figure 3.9 shows the posterior distribution of the conditional choice probabilities for a much smaller sample of 100 but with the same uniform distribution of x. A sample size of 100 is remarkably small for any non-parametric method even for one involving continuous variables. Here there is a discrete dependent variable and only 100 observations. In spite of these challenging conditions, the non-parametric method advanced here works quite well. However, if a non-uniform distribution of x is used, then the results are quite different and the posterior properly reflects the lack of information. Figure 3.10 shows a sample of 500 from the same model but with the distribution of x concentrated around 4. As might be expected, the posterior interval for the conditional choice probability is very tight around $x = 4$ but opens to the entire interval between 0 and 1 for x values not in the vicinity of 4.

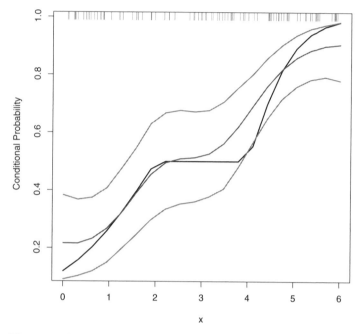

Figure 3.9. Non-parametric Choice Model: $N = 100$, x Concentrated Around 4

3.5 An Example of Expenditure Function Estimation

I will now apply the non-parametric regression approach to estimation of an expenditure function. The Homescan service of Nielsen Inc. is a national representative panel of more than 30,000 households whose purchases are tracked via scanning equipment.[2] All food and packaged goods purchases are included at the transaction level in this database. One of the

---

[2] The Kilts Center for Marketing at the Booth School of Business, University of Chicago, generously makes this data available to academic researchers.

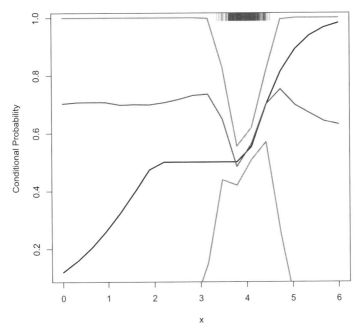

Figure 3.10. Non-parametric Choice Model: N = 100, Uniform *x*

major issues in marketing is how non-branded, Private Label (PL) products can complement nationally branded products. Private label products are important in many categories including both dry goods grocery and household products like laundry detergent. Until the availability of the complete Homescan data, very little was known about how demographic variables are related to PL consumption. In particular, there are conflicting opinions about the relationship between income and private label demand. What is known about the determinants of private label demand is from aggregate data at the store or market level. At the store or market level, there is limited variation in demographic variables and the information comes from

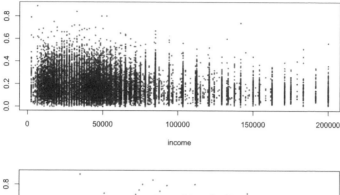

Figure 3.11. Private Label Share Vs. Income and Age

aggregate census data which has considerable measurement error. In addition, each geographic area of aggregation is characterized by an entire distribution of demographic information which is usually summarized by a simple average or median.

With the extensive household level data in the Homescan database, we should be able to determine the nature of the relationship between income and age and PL expenditure. There is no a priori reason to believe that there is a linear relationship between demographic variables and household expenditures on private label products. To illustrate the non-parametric regression method and to begin an exploration of this new data, we fit a non-parametric regression of PL expenditure share on

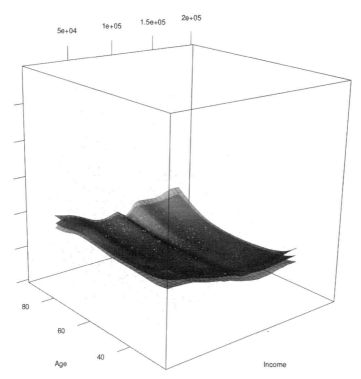

Figure 3.12. Expenditure Surface and Posterior Interval Surfaces: View from Origin

income (\$)[3] and age (years). PL expenditure share is the share of expenditure on all private label products in the year 2006. In 2006, there are 36,950 households in the Homescan database. Excluding missing values, we have a sample size of 28,717 households.

Figure 3.11 plots the PL share versus income and age. It is hard to detect a relationship between these variables. But we

---

[3]To minimize measurement errors (the questionnaire administered by Nielsen is ambiguous regarding which year the income is measured in), I take the average of 2006, 2007, and 2008 incomes.

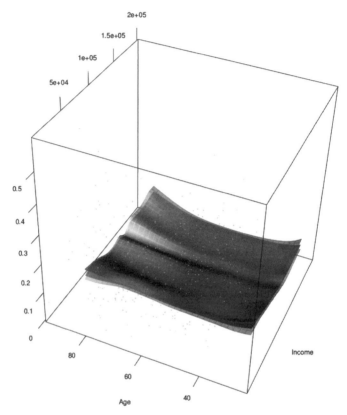

Figure 3.13. Expenditure Surface and Posterior Interval Surfaces: View from Above

note that the marginal distributions of PL share, income, and age are very dense and have no "holes" or areas of sparseness except at the very high income levels. The bivariate distribution of PL share and income or PL share and age are also dense. This means that this data set (which also has a reasonably large number of observations) is quite usable for non-parametric purposes.

Figure 3.12 shows the fitted regression or conditional mean function from a 10 component mixture of normals fit with

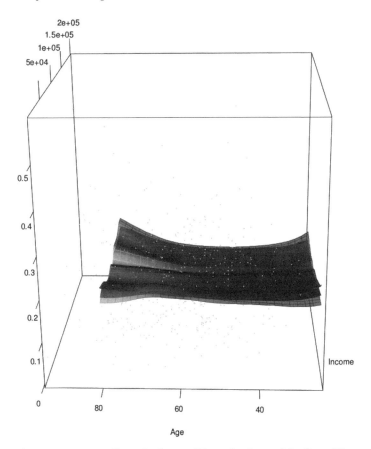

Figure 3.14. Expenditure Surface and Posterior Interval Surfaces: View from the "Age" Side

default prior settings as well as the 90 percent posterior interval. The conditional mean surface is the middle plane, and the lower (.05) and upper (.95) posterior interval boundary surfaces are the lower and upper planes, respectively. A random sample of 500 of the 28,717 observations are shown with small dots. The posterior intervals are quite tight given the large number of observations and the relatively even spread of the observations through the data. The PL share is a decreasing function of income for all values of age.

Figure 3.13 shows the expenditure surface shown from above. There is a "valley" in the surface at about the $125,000 income level. This is not due to over-fitting or sparseness in the data as the posterior intervals clearly show that evidence in the data favors the presence of this feature. This may indicate the role of a covariate not included in this analysis—for example, educational attainment. This remains as a subject for future analysis. It is important to note that it would be very difficult to uncover this feature in the data by conventional regression fits or by visualization given the large number of data points and the fact that these variables only account for about 10–15 percent of the variation in PL share.

Finally, Figure 3.14 shows the "age" side of the expenditure surface. There is modest increase in PL share between 40 and 65 years and a falling off after that point. This may reflect age-cohort differences in PL acceptance and usage.

4

Semi-parametric Approaches

As I have argued in the introduction and illustrated in Chapter 3, a fully non-parametric approach to Bayesian model inference will involve a non-trivial density estimation problem. For example, a truly non-parametric approach to regression involves all aspects of the conditional distribution of $y|x$ as the object of the modeling exercise. This makes great demands of the data, particularly for high dimensional problems. In many applications, there is insufficient data to apply a fully non-parametric approach. In these cases, we must rely on parametric methods for at least some parts of the problem. For example, single-index models in which a linear index of the x variables enters the conditional distribution of $y|x$ are one way to reduce the dimensionality of the problem. In this chapter, I will discuss Bayesian semi-parametric approaches to regression and instrumental variable estimation. A fully non-parametric Bayesian approach is available for standard regression models (see Chapter 3) but has not yet been introduced for instrumental variable models.

4.1 Semi-parametric Regression with DP Priors

Consider a linear regression model with an unspecified error distribution.

$$y_i = x_i'\beta + \varepsilon_i \tag{4.1.1}$$

If x_i and ε_i are independent, then the model can be simplified.

$$\varepsilon_i | x_i \sim \varepsilon_i$$
$$\varepsilon_i \sim iid \; F \qquad\qquad (4.1.2)$$

This rules out, at least a priori, conditional heterogeneity in the distribution of the error terms. In this situation, least squares would be consistent but not necessarily efficient. A likelihood-based semi-parametric approach would explicitly model the distribution of the error term and allow for greater efficiency, particularly with respect to the treatment of outliers. That is, a semi-parametric approach which models the distribution of the error terms would be a type of robust estimator that would be less sensitive to outliers. Outliers are viewed as observations arising from an error distribution with thicker tails than normal.

Either finite or infinite mixtures of normals models can be used to approximate the distribution of the error terms. Koop (2003) illustrates the Bayesian treatment of this model with a finite mixture of normals. Others (See, for example, Fernandez and Steel (2000) and references therein) have considered modeling the error terms as a scale mixture of normals in order to allow for outliers. In this chapter, I will develop an infinite mixture of normals approach to this problem.

The most general possible model is to allow each error term to have a possibly distinct set of normal parameters.

$$\varepsilon_i \sim N\left(\mu_i, \sigma_i^2\right)$$

A Bayesian approach requires a prior on the set of $\theta_i = \left(\mu_i, \sigma_i^2\right)$. The DP prior developed in Chapter 2 provides a simple and practical solution to this problem of assessing a prior. The DP prior "clusters" observations into one of I^* unique values of θ_i. As discussed in Chapter 2, the DP prior has both a tightness parameter and a base distribution which is indexed by a set of hyper-parameters, $\theta_i \sim G\left(\alpha, G_0\left(\lambda\right)\right)$. In Chapter 2, we developed a flexible model which adds a prior on α and also on λ. The prior on α is important and adds flexibility to the model while the prior on λ helps avoid unintentionally informative priors which put too much mass on either very "large" or very "small" normal components and hinder the flexibility of the DP prior approach. In this chapter, I will use a simpler approach of scaling the data and then directly assessing the values of λ.

The prior on α developed in Chapter 2 (2.5.3) will be used. It should be emphasized that the prior on α is assessed by considering the range of possible numbers of normal components that might be a priori reasonable. Typically, this is done in reference to the size of the data set (for example, I think it highly unlikely that models with mixtures of 20 or more components have high prior probability for a sample size of 100).

In order to retain the flexibility of the mixture of normals approach, it is important to allow both the location and scale of the mixture components to change. This means that we do not want to constrain the regression error term to have an unconditional mean of 0. This is achieved by omitting the intercept from the regression model (4.1.1) and allowing the error term to have a non-zero mean.

The DP prior used here specifies that all errors are exchangeable and, therefore, does not allow explicitly for heterogeneity such as conditional heteroskedasticity. However, the posterior distribution can reflect conditional heteroskedasticity. The posterior will group observations into common "clusters" or components that have similar error variances. If the error variance is related to the values of x then the DP posterior will cluster together observations into high and low variance components depending on their x values without assuming any given functional form for the dependence of the error term variances on x. The DP prior will serve to shrink the posteriors toward the case of homoskedasticity. However, with the fairly diffuse prior settings employed here, I do not expect a high degree of shrinkage except in situations in which the sample information is weak.

A Gibbs sampler for both the regression coefficients and the error distribution parameters can be obtained simply via data augmentation with the indicator vector, z. Given the assignment of each observation to a component, we can simply normalize the errors to have zero mean and unit variance and proceed with a standard normal conjugate prior to draw the regression coefficient vector given the normal component parameters. Given the regression coefficients, the error terms can be computed and used by the DP sampler as "data."

4.1.1 Examples of Semi-Parametric Regression with a DP Prior

I will start with an example where the error distribution is not normal and exhibits pronounced skewness. This is achieved with a translated χ^2 error term. $N = 500$ and $X \sim \text{Unif}(0, 1)$.

$$y_i = \beta_0 + \beta_1 x_i + \varepsilon_i$$

$$\varepsilon_i \sim iid \ \chi_3^2 - 3 \qquad\qquad (4.1.3)$$

The DP regression model does a good job of recovering the regression parameters as well as capturing the error structure. Figure 4.1 shows the fitted error density from the DP regression procedure using default priors (see DP prior parameterizations in 2.5.2 and 2.5.4). As noted above, I fix the λ hyper-parameters of the G_0 base distribution of the DP prior.

$$I_{min}^* = 1, \ I_{max}^* = 50, \ \omega = .8$$

$$a_\mu = .2, \ \nu = 4, \ V = 2$$

A more challenging example will be to include conditional heterogeneity. Again, $N = 500$ and $x \sim \text{Unif}(0, 1)$.

$$y_i = \beta_0 + \beta_1 x_i + \varepsilon_i \qquad\qquad (4.1.4)$$

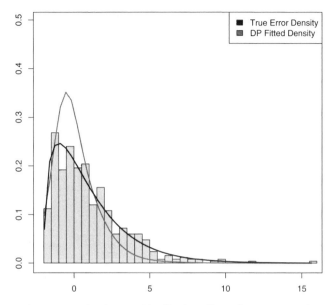

Figure 4.1. Fitted Error Distribution: Skewed Error Example

$$\varepsilon_i \sim N\left(0, \sigma_i^2\right) \tag{4.1.5}$$

$$\sigma_i^2 = \exp\left(-6.6 + 10x_i\right) \tag{4.1.6}$$

These parameter settings impart a pronounced conditional heteroskedasticity as shown in Figure 4.2. The dark line shows the true regression line and the dashed lines show $\pm 2\sigma \,|x$.

The posterior distribution of the slope parameter is shown in Figure 4.3 (the dark vertical line is the true parameter value). The top histogram shows the posterior distribution of the slope parameter, β_1, computed under the assumption that the error terms are normal and homoskedastic.[1] The bottom histogram shows the estimated posterior distribution under the DP prior.

---

[1] I use the *bayesm* routine, *runiregDP*.

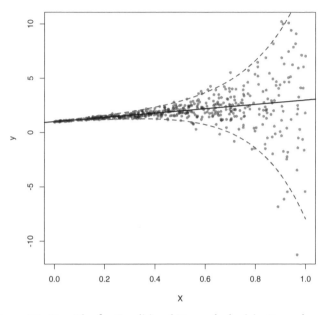

Figure 4.2. Data Plot for Conditional Heteroskedasticity Example

Although the DP does not explicitly allow for conditional heteroskedasticity, the posterior of the slope parameter clearly reflects efficiencies from exploiting the heteroskedastic structure uncovered in the data. The homoskedastic single normal component prior weights all observations with the same weight, while the DP prior does something akin to GLS—up-weighting the observations with low error variance and down-weighting those with higher variance. This is all done automatically and without formulating a conditional variance specification. Figure 4.4 plots the true values of σ_i used to simulate the data versus the posterior mean of σ_i estimated by the average of the standard deviations of the components assigned to each observation in the DP MCMC sampler (the light line is the 45 degree line). The posterior does reflect the conditional heteroskedastic structure found in the data in the sense that there is some association between the

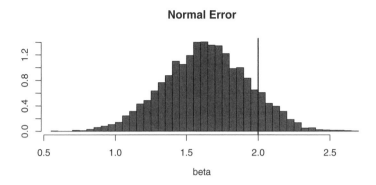

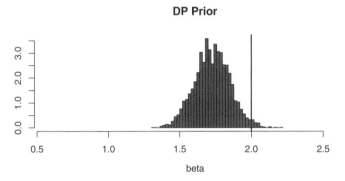

Figure 4.3. Posterior Distribution of β: DP and Normal Priors

posterior means and the true sigma values. Obviously, only a small number of "observations" (revealed only imperfectly through the realized regression errors) inform the posterior of each σ_i. The posterior reflects the dependence that the DP prior imparts by clumping or clustering observations into components with the same variance parameter. The posterior also reflects the prior in that the errors are viewed as exchangeable and, therefore, not heteroskedastic. This imparts a shrinkage toward the homoskedastic case which is quite evident in Figure 4.4. The shrinkage is most apparent for the largest values of σ_i.

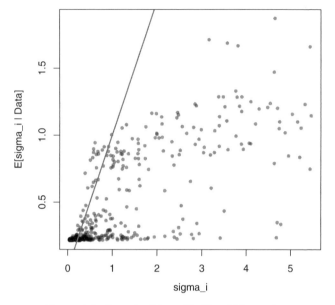

Figure 4.4. Posterior Means Vs. True Values of σ_i

4.2 Semi-parametric IV Models

Instrumental variable approaches are very important in applied empirical work in economics and marketing.[2] Consider a standard linear equation.

$$y_i = \beta x_i + w_i' \gamma + \varepsilon_i \qquad (4.2.1)$$

Some interpret this equation as representing a "structural" economic relationship between y and x. β, therefore, is interpreted as a "structural" parameter which measures the "pure" effect of changes in x on y. Often there are concerns that (4.2.1)

---

[2] This section is adapted from Conley, Hansen, McCulloch, and Rossi (2008).

is not a valid regression equation in the sense that x may be correlated or related to the error term, ε. One solution to this problem, often called the "endogeneity" problem, is to use an instrumental variable. A valid instrument is a variable that does not directly affect the left-hand side variable (i.e., is uncorrelated or independent of the error term) in the regression model but can induce variation in the right-hand side variable.

Classical econometric methods exploit this "exclusion" restriction (that is, the instrument can have no direct affect on-left-hand side variable—it is excluded from the structural equation) and propose an estimator which can be viewed as a method of moments estimator that exploits the orthogonality condition between the instrumental variable, z, and the error term, ε. Classical instrumental variable estimators are not likelihood-based in the sense that they derive solely from the orthogonality condition between the instrument and the structural equation error and are agnostic regarding any distributional assumptions. In much applied work, variables which can be logically defended as valid instruments are often only weakly related to the x variable and, for this reason, sometimes very large sets of instrumental variables have been used. Instrumental variable methods only use a portion of the variation in the x (that which is induced by the set of instruments) and, therefore, the intuition about what might be regarded as a "large" or highly informative sample that applies to standard linear regression may not apply to the instrumental variable problem. It has been well-noted that standard asymptotic approximations to the sampling distribution of instrumental variable estimators are inadequate, particularly in the "many" or "weak" instrument case.[3]

In contrast, the Bayesian treatment of the instrumental variable problem has focused on parametric versions of the IV

[3] See, for example, the discussion and references in Andrews, Moreira, and Stock (2006).

problem. Consider the linear instrumental variable problem.

$$y_i = \beta x_i + w_i' \gamma + \varepsilon_{y,i}$$
$$x_i = z_i' \delta + \varepsilon_{x,i} \tag{4.2.2}$$

Because all Bayesian methods are likelihood-based, further assumptions regarding the distribution of the error terms were introduced. Normal distributions have been used by a number of authors in Bayesian treatments of the linear IV model in (4.2.2).[4]

$$\begin{pmatrix} \varepsilon_y \\ \varepsilon_x \end{pmatrix} \sim N(0, \Sigma) \tag{4.2.3}$$

In this section, I will review a semi-parametric method that will keep the assumption of a linear structural equation but replace the assumption of normality with a general continuous distribution approximated using a DP prior.

4.2.1 The Normal Case

It is instructive to review the normal model as our DP based approximation is, after all, an extension of the normal model which allows each observation to have possibly different normal parameters. With normal errors, it is a simple matter to derive the likelihood function for the linear model. To derive the joint distribution of $y, x|z$, we simply substitute the instrumental variable equation into the structural equation. In this section, I simplify (for expositional purposes) the model in (4.2.2) by removing the exogenous variables (w) from the "structural"

---

[4]See, for example, Geweke (1996), Chao and Phillips (1998), and Hoogerheide, Kleibergen, and Dijk (2007).

equation,

$$x = \pi_x' z + v_x,$$
$$y = \pi_y' z + v_y, \tag{4.2.4}$$

with

$$\begin{pmatrix} v_x \\ v_y \end{pmatrix} \sim N(0, \Omega), \ \Omega = A\Sigma A', \ A = \begin{bmatrix} 1 & 0 \\ \beta & 1 \end{bmatrix} \tag{4.2.5}$$

and

$$\pi_x = \delta, \ \pi_y = \beta\delta.$$

Our view is that it is best to put priors directly on (β, δ, Σ) rather than on the reduced form coefficients and reduced form error covariance matrix. In particular, it would be inappropriate to assume that Ω and π_y are a priori independent since both sets of parameters depend on β. A useful starting point is to use conditionally conjugate and independent priors for the linear system.

$$\delta \sim N\left(\overline{\delta}, A_\delta^{-1}\right), \ \beta \sim N\left(\overline{\beta}, A_\beta^{-1}\right), \ \Sigma \sim IW(\nu, V) \tag{4.2.6}$$

It is easy to define a Gibbs sampler for the system (4.2.2) and (4.2.6). The Gibbs sampler contains three sets of conditional posterior distributions.

$$\beta | \delta, \Sigma, y, x, Z$$
$$\delta | \beta, \Sigma, y, x, Z \tag{4.2.7}$$
$$\Sigma | \beta, \delta, y, x, Z$$

The intuition for the Gibbs sampler is the same intuition that motivates the "endogeneity" problem in the first place. We know that the linear structural equation for y is not a valid regression equation because the error term has a non-zero mean which depends on x. However, the first distribution in the Gibbs sampler conditions on δ, which means that we can "observe" ϵ_x and the conditional distribution of $\epsilon_y|\epsilon_x$ can be derived. This distribution can be used to convert the structural equation into an equation with a N(0, 1) error term.

$$\left(y - \frac{\sigma_{xy}}{\sigma_x^2}\epsilon_x\right) = \beta x + u, \ u \sim \text{N}\left(0, \sigma_y^2 - \frac{\sigma_{xy}}{\sigma_x^2}\right) \qquad (4.2.8)$$

Dividing through by σ_u converts the first Gibbs sampler draw into Bayes regression with a unit variance error term. The second conditional in the Gibbs sampler is simply a restricted two-variate regression which can be achieved easily by "doubling" the observations with rows of z_i and βz_i. Given β, we can re-write the reduced form.

$$z = z'\delta + \varepsilon$$

$$\tilde{y} = \left(\frac{y - w'\gamma}{\beta}\right) = z'\delta + \left(\varepsilon_x + \frac{\varepsilon_y}{\beta}\right) \qquad (4.2.9)$$

We can transform (4.2.9) by the appropriate Cholesky root to transform the problem into a "stacked" Bayesian regression with N(0, I_2) errors. The final Gibbs sampler conditional is a standard IW draw. This Gibbs sampler is implemented in the *bayesm* routine, `rivGibbs`.

4.2.2 Mixture of Normals and DP Errors

A legitimate concern about the Bayesian "IV" procedure is the additional specification of normal error terms in both the structural and "first-stage" instrument equation. This is not required for classical IV estimators. Usually, this is a concern about the possible inconsistency of an estimator based on a mis-specified likelihood. Equally important, but rarely appreciated, is the possibility for improved inference if it is possible to detect and model the non-normal structure in the error term. For example, suppose that the error terms are a mixture of a normal error with small variance and another normal with a very large variance. The single component normal error model may be sensitive to the outliers and will treat the outlying error terms inefficiently. In principle, it should be possible to detect and down-weight observations that appear to have errors drawn from the outlying component. The classical IV approach sacrifices efficiency for the sake of consistency. In a semi-parametric Bayesian approach, it is theoretically possible to be robust to mis-specification while not reducing efficiency. That is, in the normal case we might not lose much or any efficiency but in the non-normal case we exploit the structure and construct an efficient procedure.

A logical place to start a departure from normality is the mixture of normals model. Intuitively, if we condition on the latent indicator variable for the membership in normal components, we should be able to reuse the Gibbs sampler for the normal model as we can adjust and properly normalize for different variances. If mixtures of normals are used to approximate the unknown bi-variate density of the error terms in the linear system (4.2.2), the means of each component must not be fixed at zero. A mixture of normals with zero or equal means is simply a scale mixture and cannot approximate skewed or multi-modal distributions. For this reason, the intercepts are removed from both the instrument and structural equations.

$$x = Z\delta + \epsilon_x^*$$
$$y = \beta x + \epsilon_y^*$$
$$\begin{pmatrix} \epsilon_x^* \\ \epsilon_y^* \end{pmatrix} \sim N(\mu_z, \Sigma_z) \qquad (4.2.10)$$
$$z \sim MN(\pi)$$

To complete this model, we need to put a prior over the number of normal components. In the standard finite mixture of normals, we assume that there are up to K possible unique values. A DP prior can be used which puts positive prior probability on up to N unique components. The DP process, $G(\alpha, G_0)$, defines this prior. A prior on α is assessed using the parameterization introduced in (2.5.3). The prior on α is chosen to put prior mass on values of the number of unique components from 1 to at least 10 or more, though we note that the DP prior puts positive prior probability on up to N (the sample size) unique values or components.

Instead of assessing further priors on the hyper-parameters of the base prior distribution, G_0 (see 2.5.4), Conley, Hansen, McCulloch, and Rossi (2008) directly assess these prior parameters. The following parameterization of G_0 is used.

$$G_0: \ \mu|\Sigma \sim N(\bar{\mu}, a^{-1}\Sigma), \ \Sigma \sim IW(v, V) \qquad (4.2.11)$$

To assess G_0, we center and scale both y and x. For centered and scaled dependent variables, we would expect that the bivariate distribution of the errors terms is concentrated on the region, $[-2, 2] \times [-2, 2]$. In order to achieve full flexibility, we want the possibility of locating normal components over a "wide" range of values and with a reasonable range of small and large variances.

Given our standardization, $\bar{\mu} = 0$ is an obvious choice and $V = vI$ where I is the identity matrix is a reasonable choice. The prior on μ governs the location of the normal components and the prior on Σ influences both the spread of the μ values as well as the shape of each normal component. If a large number of components are used, it is unlikely to matter much what shapes are a priori most probable (i.e., whether the normal components are correlated or uncorrelated normals). However, with small numbers of components, this may matter. Our assessment procedure allows for a relatively large (potential) number of normal components. This means that the prior on μ determines the implied prior of the error distribution. Since the normal components are spread over the error space evenly in all quadrants, our prior specification weakly favors independence of the errors. Thus, in the case of data sets with little information, the Bayes estimators proposed here will "shrink" toward the case of no endogeneity.

To see this, consider the predictive distribution of the error terms under the Dirichlet Process prior. Recall that the marginal distribution of $\theta_i = (\mu_i, \Sigma_i)$ is $G_0(\lambda)$. Therefore, the predictive distribution, $p(\varepsilon) = \int p(\varepsilon|\theta) \, dG_0(\theta|\lambda)$, is a multivariate student t with location matrix, V. Since we have taken the parameterization, $V = vI_2$, this distribution has zero correlation between the two errors (there is dependence but only in the scale). If we put in a non-diagonal value for V, this could be changed. The problem, however, is that we rarely have a priori information about the sign and magnitude of the error dependence. Thus, a diagonal V is a reasonable default choice.

We now have three scalar (v, v, a) quantities to choose. To make these choices we must think about what kinds of normal distributions are possible for the ε_i. The largest errors would occur when δ, β, and γ are zero so that the ε_i are like the (x_i, y_i). Thus, we need the priors to be spread out enough for μ to cover possible (x_i, y_i) values and Σ to be large enough to be consistent with the observed variation of (x_i, y_i).

To make the choice of (v, v, a) more intuitive, consider the implied marginals for $\sigma = \sqrt{\sigma_{xx}}$ and μ.[5] We assess intervals for these marginals and then compute the corresponding values for (v, v, a). Define the intervals as follows:

$$\Pr[c_1 < \sigma] = \Pr[\sigma < c_2] = \kappa/2 \qquad (4.2.12)$$

$$\Pr[-c_3 < \mu < c_3] = 1 - \kappa$$

Given κ, choosing (v, v, a) is equivalent to choosing c_1, c_2, and c_3. For example, if we use $\kappa = .2$, $c_1 = .25$, $c_2 = 3.25$, and $c_3 = 10$, this gives $v = 2.004$, $v = .17$, and $a = .016$. These values we term our "default" prior and are used in the sampling

---

[5]The distributions of each of the error terms are treated symmetrically.

experiments and empirical examples. While these choices may seem to be "too" spread out given our choice of standardization, the goal is to be as diffuse as possible without allowing absurd choices. If the resulting posteriors are very sensitive to these prior choices, then we would have a problem. However, we will see in our examples that this is not the case. The marginal distributions required to evaluate (4.2.12) are

$$\sigma \sim \chi_\nu^2, \; \mu \sim \sqrt{\frac{\nu}{a}(\nu-1)} t_{\nu-1}. \qquad (4.2.13)$$

In the Polya Urn method for drawing from the posterior distribution in DP models, $\theta_i = (\mu_i, \Sigma_i)$ components are drawn for each observation. However, these values are clustered to a smaller number of unique values. The indicator variable can be formed from the set of draws of the errors distribution parameters and the set of unique values. This means that we can form a Gibbs sampler for the linear IV model with a DP prior on the errors from the following steps:

$$\beta | \delta, z, \{\theta_i\}, x, y, Z$$
$$\delta | \beta, z, \{\theta_i\}, x, y, Z$$
$$\{\theta_i\} | \beta, \delta, x, y, Z \qquad (4.2.14)$$
$$\alpha | I^*$$

Given a set of draws of $\{\theta_i, i = 1, \ldots, N\}$, we can define the I^* unique values as $\{\theta_j^*, j = 1, \ldots, I^*\}$. The indicator vector, z, is defined by $z_i = j$ if $\theta_i = \theta_j^*$. The draws of β and δ in (4.2.14) are basically the same as for the normal model except that adjustments must be made for the means of the error terms and there are different means and variance terms depending on which unique value is associated with each observation. The *bayesm* routine, rivDP, implements the full Gibbs sampler in (4.2.14).

4.2.3 Sampling Experiments

The Bayesian semi-parametric procedure outlined in section 4.2.2 can be evaluated by comparison with other Bayesian procedures based on normally distributed equation errors or by comparison with classical procedures. Comparison with a Bayesian procedure based on normal errors is straightforward as both procedures are in the same inference framework. Comparison with classical procedures is somewhat more complicated as classical methods often draw a distinction between what is termed an "estimation problem" and an "inference problem." A variety of *k*-class estimation procedures have been proposed for the linear structural equations problem, but the recent classical literature has focused on improved methods of inference. Inference is often viewed as synonymous with the construction of confidence intervals with correct coverage probabilities. In this section, we discuss simulation experiments designed to compare the sampling properties of our Bayesian semi-parametric procedure with those of alternative Bayes and classical procedures.

Experimental Design

We consider the linear structural equation model with one endogenous right hand side variable. The simplicity of this case will allow us to explore the parameter space thoroughly and focus on the effects of departures from normality which we regard as our key contribution. Moreover, this model is empirically relevant. A survey (Chernozhukov and Hansen (2008)) of the leading journals in economics (*QJE/AER/JPE*) in the period 1996–2004 produced 129 articles using linear structural equation models of which 89 had only one endogenous right-hand side variable. It appears, therefore, that the canonical use of instrumental variable methods is to allay concerns about endogeneity in regression models with a small number of potentially endogenous right-hand side variables. The model considered in our simulation experiments is given by

$$x = z'(\iota\delta) + \varepsilon_x, \tag{4.2.15}$$

$$y = \beta x + \varepsilon_y. \tag{4.2.16}$$

ι is a vector of k ones. Throughout we assume that $\beta = 1$. Since the classical literature considers both the case of "many" instruments and weak instruments, we will consider the case in which z is of dimension $k = 10$. Each element of the z vector is generated as *iid* Unif$(0, 1)$ and the z are redrawn for each simulation replicate. We specify both a normal and log-normal distribution of the error terms in 4.2.15.

$$\begin{pmatrix} \varepsilon_x \\ \varepsilon_y \end{pmatrix} \sim N(0, \Sigma) \tag{4.2.17}$$

or

$$\begin{pmatrix} \varepsilon_x \\ \varepsilon_y \end{pmatrix} = cv, \quad v \sim lnN(0, s\,\Sigma). \tag{4.2.18}$$

$\Sigma = \begin{bmatrix} 1 & .6 \\ .6 & 1 \end{bmatrix}$, $s = .6$, and c is taken so that the interquartile range of log-normal variables is the same as the normal distribution in the base case. The idea is to create skewed errors without an excessive number of positive outliers.

The value of δ is chosen to reflect various degrees of strength of instruments from "weak" to "strong" settings. In the classical literature, the F statistic from the first stage regression or the concentration parameter, kF, is used to assess the strength of the instruments. It is also possible to compute the population R-squared implied by a particular choice of δ.

$$\rho^2 = \frac{\frac{1}{12}\delta^2 k}{\frac{1}{12}\delta^2 k + \sigma_{11}} \tag{4.2.19}$$

We chose two[6] values of δ, (.5, 1.5), hereafter referred to as "weak" or "strong." For the normal distribution case, these correspond to population R-squared values of (.17, .65). These are the approximate quartiles of the empirical distributions of R-squared found in our literature search.[7]

We use a sample size, N, of 100. For each of 400 replications, we draw errors from the two error distributions specified in (5.2.17) and (5.2.18). We repeat this process for each of two values of δ, resulting in a sampling experiment with four cells. Our weak instrument cells contain a substantial number of data sets with first-stage R-squares below 10 percent or concentration parameters below 10.

For each of our generated data sets, we will compute a Bayesian 95 percent Credibility Interval and the posterior mean using our semi-parametric procedures. We will compare these intervals and estimates to Bayesian estimates using normal errors, k-class estimators, and intervals constructed using standard asymptotics, many instrument asymptotics, and weak instrument asymptotics.

Results on Interval Coverage

Interval estimates were constructed from the following set of procedures: Standard k-class asymptotics—OLS, TSLS, LIML, F1 (Fuller estimator):

- Many instrument asymptotics: LIML-M and F1-M (Bekker (1994))
- Weak instrument asymptotics: K (Kleibergen (2002)), J (Kleibergen (2007)), and CLR (Moreira (2003))

---

[6]Conley, Hansen, McCulloch, and Rossi (2008) also consider a "moderate" strength case, which I omit in the interest of brevity.

[7]Some may object to our characterization of .17 with 10 instruments as a case of weak instruments. The implied p-value for the F statistic in the weak instrument case with 10 instruments and 100 observations given the population R-squared is .068.

- Bayesian: Bayes-NP (Bayesian procedure assuming
  normal errors) and Bayes-DP (Bayesian procedure using
  DP prior for error distribution)

All classical intervals were constructed with a nominal confidence level of .95. The Bayesian intervals were constructed
using the .025 and .975 quantiles of the simulated posterior
distribution. Some of the weak instrument procedures produce
empty intervals and infinite length intervals. In addition to
reporting the actual coverage probabilities, we report an Interval
Measure (IM), and the number of infinite and empty intervals.
The interval coverage measure is defined by

$$\text{IM} = \frac{1}{U - L} \int_{L}^{U} |x - \beta| dx. \qquad (4.2.20)$$

This measure can be interpreted as the expected distance of a
random variable, distributed uniformly on (L, U) from the true
value, β.

We use the "default" prior settings for λ given above. In the
prior for α, we assess $(\underline{\alpha}, \overline{\alpha})$ corresponding to modes of one and
eight, respectively, and $\omega = .8$.

Table 4.1 shows the results for the weak instrument case. The
best coverage is obtained by the K, CLR, and many instrument
methods that achieve actual coverage close to .95 for both the
normal and non-normal cases. The Bayes procedures provide
intervals whose coverage is below the specified level. For normal
errors, the Bayes-NP and Bayes-DP methods produce very similar
coverages. The coverage of the Bayes-NP procedure degrades
under log-normal errors while the Bayes-DP procedure has
coverage close to .95.

However, coverage is not the only metric by which performance can be judged. The K, J, and CLR methods produce a
substantial number of infinite length intervals. In particular,
the CLR method produces infinite length intervals about 40
percent of the time for the case of log-normal errors. If the data

Table 4.1. Comparison of Procedures: Weak Instrument Case

Procedure	Normal				Log-normal			
	Coverage	IM	Infinite	Empty	Coverage	IM	Infinite	Empty
OLS	0.0	.5	0	0	0.0	.5	0	0
TSLS	.75	.27	0	0	.69	.37	0	0
LIML	.92	.36	0	0	.89	.64	0	0
F1	.92	.32	0	0	.89	.49	0	0
LIML-M	.94	.40	0	0	.93	.75	0	0
F1-M	.93	.35	0	0	.92	.61	0	0
K	.94	1.38	118	0	.95	2.12	270	0
J	.89	.84	31	16	.93	1.61	173	6
CLR	.92	.75	27	0	.96	1.58	168	0
Bayes-NP	.84	.26	0	0	.79	.35	0	0
Bayes-DP	.83	.26	0	0	.91	.18	0	0

set provides no information about the value of the structural parameter, then one might justify producing an infinite length interval. In this case, it is unlikely that over 40 percent of the weak instruments simulation data sets have insufficient information to make inferences at the 95 percent confidence level. It appears that the log-normal errors create difficulties for the weak instrument procedures.

The interval measure provides a measure of both the position and length of each interval. The Bayes-DP procedure provides IM values dramatically smaller than all other procedures (particularly the weak instrument methods). Note that infinite intervals are truncated to (-5, 5). It is not only the infinite intervals that create the large values of the interval measure. For example, the F1-M procedure (which produces only finite intervals) has an IM which is three times the size of the Bayes-DP procedure in the log-normal case. All of the classical procedures have IM values which are substantially larger in the case of log-normal errors. The smaller size of the Bayes-DP intervals is not simply that they have lower coverage rates. In the case of log-normal errors, the DP procedure has a coverage rate of .91 yet an interval measure less than one-third of any instrument procedure and less than one eighth of any weak instrument method. The F1-M procedure has virtually the same coverage rate but produces intervals 3 times longer.

The Bayes-DP procedure captures and exploits the log-normal errors and provides a lower interval measure in the case of log-normal errors. The log-normal distribution has many large positive outliers which are down-weighted by the DP. The remaining errors are small. The idea is that if you can devote normal components to the outliers and errors clustered near zero then you will obtain superior interval estimates.

These same qualitative results hold even for the strong instrument strength cases as presented in Table 4.2. There are still some cases of infinite length and empty intervals for some of the weak-instrument asymptotic approximations. The Bayesian procedures now provide coverage values close to the nominal

Table 4.2. Comparison of Procedures: Strong Instrument Case

Procedure	Normal				Log-normal			
	Coverage	IM	Infinite	Empty	Coverage	IM	Infinite	Empty
OLS	.06	.2	0	0	.03	.29	0	0
TSLS	.92	.09	0	0	.90	.14	0	0
LIML	.96	.10	0	0	.96	.15	0	0
F1	.95	.10	0	0	.96	.14	0	0
LIML-M	.95	.10	0	0	.96	.15	0	0
F1-M	.95	.09	0	0	.96	.14	0	0
K	.94	.15	9	0	.93	.45	41	0
J	.92	.17	0	11	.90	.24	0	13
CLR	.94	.10	0	0	.95	.16	0	0
Bayes-NP	.94	.09	0	0	.93	.13	0	0
Bayes-DP	.92	.09	0	0	.96	.07	0	0

levels but still provide intervals that are much smaller. In the "strong" case, the Bayes-DP method provides coverage that is almost exactly correct but with intervals that are one-half to one-third the length of all other classical procedures. We note that the "strong" instrument case is calibrated to a first stage F statistic that is roughly equal to the 75th percentile of the empirical studies surveyed in the literature. The weak instrument cell is calibrated so that the population R-squared and F correspond to the 25th percentile of the survey of empirical work. However, sampling variation results in many simulated data sets in this cell with much weaker instruments than appear in published work. It is instructive to compare the Bayes-NP which assumes normal errors with the Bayes-DP procedure that does not. Both procedures produce nearly identical intervals for the case of normal errors, while the Bayes-DP procedure produces much smaller intervals with less "bias" in location for the case of log-normal errors.

The difference between the various methods is best illustrated graphically. Figure 4.5 presents the Bayes-DP and CLR intervals for the first 50 simulated data sets in the weak instrument, log-normal case. The Bayes-DP intervals are represented by dark lines and the CLR intervals by light lines. Infinite intervals are indicated by dashed lines. The true parameter value is drawn on the figure as a vertical line. The Bayes intervals are dramatically smaller but exhibit a positive "bias" in location. Again, the Bayesian non-parametric procedure discovers and exploits the non-normality to produce dramatically smaller intervals. Much the same is true for the comparison of the Bayes-DP procedure to F1-M displayed in Figure 4.6.

Results on Estimation Performance

Our method can uncover and exploit structure in the data which opens the possibility of greater efficiency in estimation without a consistency-efficiency trade-off. For these reasons, we will also investigate the estimation performance of our Bayesian method. We compare this to the standard OLS, TSLS, LIML, and F1

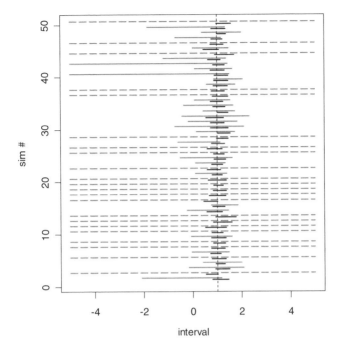

Figure 4.5. Comparison of Bayes-DP and CLR Intervals: Weak Instruments and Log-normal Errors

methods as well as to the Bayes-NP method which assumes normal errors.

Table 4.3 provides standard metrics including RMSE, Median Bias, and the Interquartile Range (IQR) of the sampling distribution. The Bayes-DP method dominates on RMSE and IQR across all cells of the experimental design. F1 and LIML have lower median bias than Bayes estimates under normal errors. In particular, it is noteworthy that the presence of log-normal errors dramatically worsens the performance of TSLS as measured by both bias and RMSE. The Bayes-DP method exploits the non-normality and produces estimates with much smaller RMSE and low bias (comparable to F1). These results are not driven by

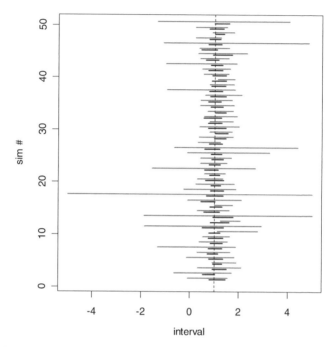

Figure 4.6. Comparison of Bayes-DP and F1-M Intervals: Weak Instruments and Log-normal Errors

outliers in the sampling distribution as comparison of the IQR measure reveals.

Prior Sensitivity

Using the approach outlined in section 4.2.2, we assessed a very diffuse prior. The key components of the DP prior are the settings of the hyper-parameters, λ, and the prior on the Dirichlet Process "tightness" parameter, α. The prior on α influences the number of unique values while λ influences the size and shape of the components drawn.

Table 4.3. Performance of Estimators

Instrument strength	Estimator	Normal			Log-normal		
		RMSE	Median bias	IQR	RMSE	Median bias	IQR
Weak	OLS	.50	.50	.09	.53	.48	.19
	TSLS	.26	.20	.22	.37	.28	.31
	LIML	.36	.02	.34	.47	.01	.57
	F1	.29	.05	.32	.43	.09	.48
	Bayes-NP	.24	.17	.23	.35	.26	.30
	Bayes-DP	.24	.18	.22	.16	.09	.16
Strong	OLS	.21	.21	.07	.31	.28	.13
	TSLS	.08	.02	.09	.12	.06	.13
	LIML	.08	-.01	.10	.12	.01	.16
	F1	.08	0	.10	.12	.01	.15
	Bayes-NP	.07	.02	.09	.11	.04	.13
	Bayes-DP	.07	.02	.09	.05	.01	.07

For the prior on α, we choose $\underline{\alpha}, \overline{\alpha}$ corresponding to a mode of I^* of 1 and 8, respectively, and set the prior power parameter, $\omega = .8$. This provides a prior which implies a distribution of I^* that puts substantial mass on values between 1 and 8, but with a long tail. This corresponds to our view that with only 100 observations, it would be foolish to attempt models with more than 10 normal components. However, unlike classical procedures, the Bayesian procedure computes Bayes Factors for the addition of new components. New components are not added unless the fit/parameters trade-off is very favorable. We experimented with data simulated under weak instruments with both normal and log-normal errors. We selected a wide range of α values and examined the resulting inference for the structural parameter, β. We found that inference was very insensitive to the choice of α.

For the λ settings, we choose a "default" setting which implies a very diffuse prior on each $\theta_i = (\mu_i, \Sigma_i)$.

$$\Pr[.25 < \sigma < 3.25] = .8 \text{ and } \Pr[-10 < \mu < 10] = .8$$

These imply prior settings

$$\Sigma \sim \text{IW}(2.004, .17I_2),$$

$$\mu|\Sigma \sim \text{N}\left(0, \frac{1}{.016}\Sigma\right). \qquad (4.2.21)$$

For comparison purposes, we consider two other prior settings.

Alternative 1: The first is chosen to be less diffuse than our "default" setting.

$$\Pr[.5 < \sigma < 3] = .9 \text{ and } \Pr[-5 < \mu < 5] = .9$$

Table 4.4. Prior Sensitivity Analysis

Instrument strength	Prior	Normal		Log-normal	
	Prior	Coverage	IM	Coverage	IM
Weak	Default	.83	.26	.91	.18
	Alt 1	.74	.26	.85	.23
	Alt 2	.74	.26	.86	.21
Strong	Default	.92	.09	.96	.07
	Alt 1	.92	.09	.94	.09
	Alt 2	.92	.09	.95	.08

and associated settings

$$\Sigma \sim \text{IW}(3.4, 1.7I_2),$$

$$\mu | \Sigma \sim \text{N}\left(0, \frac{1}{.2}\Sigma\right). \qquad (4.2.22)$$

Alternative 2: The second is less diffuse and is suggested by "standard" natural conjugate prior setting used in many Bayesian analyses of the multivariate normal problem.

$$\Pr[.4 < \sigma < 1.31] = .8 \text{ and } Pr[-.95 < \mu < .95] = .8$$

and associated settings

$$\Sigma \sim IW((4, I_2),$$

$$\mu | \Sigma \sim N\left(0, \frac{1}{.1}\Sigma\right). \qquad (4.2.23)$$

Table 4.4 provides evidence on the sensitivity of coverage and the interval measure to the prior settings. The interval measure is completely insensitive to the prior settings for all six experimental cells. The coverage probability is slightly better with the "default" prior. These simulations support our view that a diffuse proper prior will provide excellent performance without

additional tuning. We note that the view that the settings in
(4.2.21) are diffuse depends critically on the fact that we have
rescaled both y and x to have unit standard deviation and zero
mean. This allows us to take the view that the errors are on a
standard deviation scale and are unlikely to take on values that
are extremely large such as 20 or more.

Conclusions from Sampling Experiments

Sampling experiments illustrate the value of our approach by
comparison with leading large sample approximation methods
in the weak and many-instrument literature. In the weak instrument case, the coverage rates of our procedure are 4 to 12
percentage points lower than the nominal 95 percent rate. We
find that the weak instrument procedures produce very long
intervals, especially in the case of non-normal errors. In our
view, very long intervals with correct coverage is not the answer
that most applied researchers are seeking. Moreover, coverage is
not necessarily the most appropriate metric for assessing interval
estimation performance. Our Bayes intervals have somewhat
lower coverage only for the weak instrument cases. Even then,
the intervals are located "close" to the parameter values relative
to other methods. That is, when we miss, we don't miss by much.

The Bayesian semi-parametric procedure produces credibility
regions which are dramatically shorter than confidence intervals
based on the weak instrument asymptotics. The shorter intervals
from our method are produced by more efficient use of sample
information. The RMSE of our semi-parametric Bayes estimator
is much smaller than classical IV methods,[8] especially in the case
of non-normal errors. A Bayesian method that assumes normal
errors produces misleading and inaccurate inference under non-
normality and about the same answers as our non-parametric

---

[8]With lower median bias than TSLS. Our Bayesian-DP intervals have
comparable median bias as LIMI and F1 in all cases considered except in the
case of very weak instruments and normal errors.

method under normality. It appears, then, our non-parametric Bayesian method dominates Bayesian methods based on normal errors and may be preferable to methods from the recent weak instruments literature if the investigator is willing to trade off lower coverage for dramatically smaller intervals.

4.2.4 Empirical Examples

We consider two empirical examples of the application of our methods. We include examples with small and moderately large numbers of observations.

Acemoglu

The first example is due to Acemoglu, Johnson, and Robinson (2001) who consider the relationship between GDP per capita and a measure of the risk of expropriation. To solve the endogeneity problem, European settler mortality is used as an instrument. In former colonies with high settler mortality, Acemoglu and co-authors argue that Europeans could not settle and, therefore, set up more extractive institutions. We consider a specification which is a structural equation with log GDP related to Average Protection Against Expropriation Risk (APER), latitude, and continent dummies along with a first stage regression of APER on log European Settler Mortality and the same covariates as in the structural equation. The incremental R-squared from the addition of the instrument is .02 with a partial F statistic of 2.25 on 1 and 61 degrees of freedom. The least squares coefficient on APER is .42 while the TSLS coefficient is 1.41 in this specification. The conventional asymptotics yield a 95 percent C.I. of $(-.015, 2.97)$ with $N = 64$. The CLR procedure provides a disjoint and infinite C.I., suggesting that conventional asymptotics are not useful here due to a weak instrument problem.

Both the conventional and CLR-based confidence intervals extend over a very wide range of values. This motivates an interest in methods with greater efficiency. This small data set (64 observations) will also stress test our Bayesian method with a Dirichlet prior. Figure 4.7 shows the posterior distribution using normal errors (top panel) with Dirichlet prior. We use the "default" prior settings discussed in section 4.2.3. The Bayes 95 percent credibility interval is drawn on the horizontal axis with light brackets. The interval for the normal error model is (.13, 1.68) and the interval for the DP prior model is (.05,1.2). The inferences from a Bayesian procedure with normal errors are not too different from conventional TSLS estimates. However, the interval derived from the Bayes-DP model is considerably shorter and is located nearer the least squares estimates. Figure 4.8 shows the fitted density of the errors constructed as the posterior mean of the error density. This density displays some aspects of non-normality with diamond-shaped contours.

Card Example

Card (1995) considers the time-honored question of the returns to education, exploiting geographic proximity to two-and four-year colleges as an instrument. He argues that proximity to colleges is a source of exogenous changes in the cost of obtaining an education which will cause some to pursue a college education when they might not otherwise do so. The basic structural equation relates log of wage to education (years), experience, experience-squared, a black dummy variable, an indicator for residing in a standard metropolitan statistical area, a South indicator variable, and various regional indicators. The first stage is a regression of education on two indicators for proximity to two-and four-year colleges and the same covariates as in the structural equation. The incremental R-squared from the addition of the instruments to the first state is .0052 with corresponding F of 7.89 on 2 and 2993 degrees of freedom. OLS estimates of the return to education are around .07 while the TSLS estimates are much higher, around .157 with a standard error of .052, N = 3010. The LIML estimate is .164 with a standard error of .055. To our view, these returns of 14 percent per year of

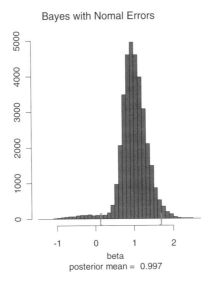

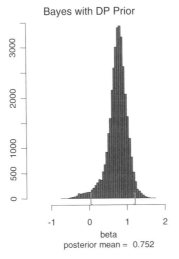

Figure 4.7. Posterior Distribution of the Structural Coefficient: Acemoglu Data

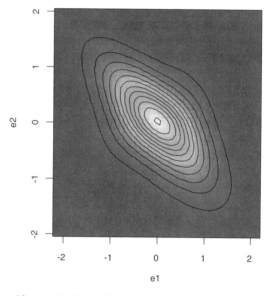

Figure 4.8. Fitted Error Density: Acemoglu Data

education seem high. However, even with more than three thousand observations, the confidence interval for the returns on education is very large.

Figure 4.9 shows the posterior distributions assuming normal errors and using the DP prior. The 95 percent posterior credibility regions are denoted by light brackets. For the normal error case, the interval is (.058, .34) while it is (.031, .17) for the DP prior model. We use the "default" prior settings for λ and values of corresponding to modes of 1 and 30, respectively. As in the Acemoglu data, the normality assumption makes a difference. With the DP prior, the posterior distribution is much tighter and centered on a lower rate of return to education. There is less "endogeneity" bias if one allows for a more flexible error distribution. Figure 4.10 shows the fitted density from the DP prior procedure (bottom panel) as well as the predictive density of the errors from a Bayesian procedure that assumes the error

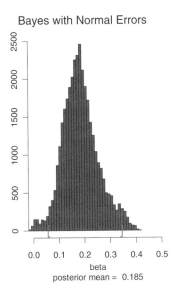

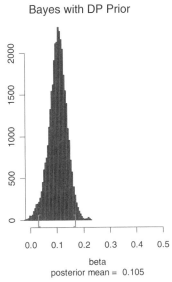

Figure 4.9. Posterior Distribution of the Structural Coefficient: Card Data

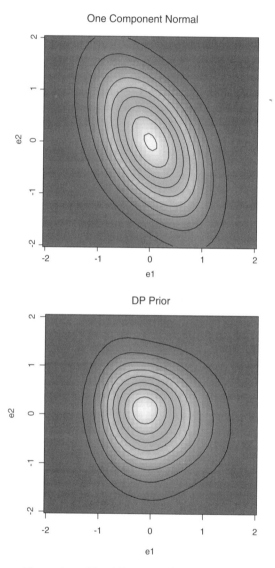

Figure 4.10. Fitted Error Density: Card Data

terms are normal (top panel). There are marked differences between these densities. The normal error model concludes that there is substantial negative correlation in the errors, while the Bayesian non-parametric model yields a distribution with pronounced skewness and non-elliptical contours, showing little dependence. It is possible that outlying observations are driving this result. In any event, the assumption of normality has substantial implications for the inference about the structural quantity. These examples illustrate that our Bayesian procedures give reasonable results for a wide range of sample sizes, and it matters whether or not one specifies a normal distribution of the error terms. Moreover, it appears that the Bayesian non-parametric model is capable of discovering and exploiting structure in the data, resulting in tighter posterior distributions.

5

Random Coefficient Models

5.1 Introduction

Random coefficient models are frequently applied in a panel data setting in which the data consist of a set of cross-sectional units observed over time. For example, the Homescan data available from the Kilts Center of Marketing at the University of Chicago's Booth School of Business tracks the purchases of some 30,000–60,000 households over a broad set of products. Each household is in the panel for a limited length of time, varying from less than one year to more than six years. In many applications, only a small and relatively homogenous set of products are considered. In these situations, there is a large set of panelists some of whom have only very short purchase histories. Thus, the essential characteristics of many panel data sets is a large number of cross-sectional units (N) and a small number of observations per unit (T). Given these types of data structures, a fully non-parametric approach to modeling panel data is not practical. For a given cross-sectional unit, a parametric model is required as the number of observations for any given unit is small. However, there are often a very large number of cross-sectional units which allows for the possibility of the use of non-parametric methods to model the variation of parameters across units. Of course, this requires a view that there is some sort of homogeneity among the panelists as a random coefficient model assumes that panel-level parameters are drawn from some common distribution. Typically, the random coefficient models used in the econometric literature are parametric models.

Parametric random coefficient models have been very popular in both the marketing and econometric literatures.[1] In classical applications, random coefficient models are applied only to a subset of model coefficients, for example, only to the intercepts in a regression model, due to limitations of classical simulation methods of inference. In the Bayesian approach,[2] there is no need to constrain the set of random coefficients to a subset of model coefficients, indeed it is often easier to allow all model coefficients to vary over the cross-sectional units in the panel. Finally, classical methods based on optimization often have great difficulty handling a random coefficient model with a full-covariance structure and, for this reason, restrict the covariance structure to eliminate parameters which prove to be difficult to estimate using optimization-based estimators. However, the full Bayesian approach uses prior information and is implemented via simulation-based methods, freeing the researcher from some of the constraints of classical methods.

The computational advantages of a fully Bayesian random coefficient approach (see, for example, Chapter 5 of Rossi, Allenby, and McCulloch (2005)) allow for more routine application of random coefficient models, particularly in non-linear model contexts. However, this does not free the random coefficient approach from criticism. One important drawback is that the model assumes a particular parametric form of the random coefficient distribution (usually a normal form).

We can formulate a random coefficient model as follows:

$$y_{jt}|x_{jt}, \theta_j \qquad (5.1.1)$$

$$\theta_j|\tau \qquad (5.1.2)$$

$$\tau|h \qquad (5.1.3)$$

---

[1] The so-called "BLP" model of Berry, Levinsohn, and Pakes (1995) is a random coefficient logit model and has been widely applied in many different contexts in the I/O and marketing literature.

[2] See, for example, Rossi and Allenby (2011).

5.1.1–5.1.3 specify a random coefficient model as a hierarchical model via a sequence of conditional distributions. The first distribution (5.1.1) is the model for the observations. For example, the first stage could be a linear regression of y on x or a discrete choice model. The second stage is the random coefficient model which specifies a particular parametric distribution indexed by τ. In the final stage (5.1.3), a prior is put over the parameters of the parametric model and indexed by the hyperparameter, h. Fully Bayesian inference for this model is usually accomplished via a hybrid MCMC method.

In many applications, the "random coefficient" portion of this model uses an unrestricted multivariate normal model. While the normal distribution is flexible, there are many obvious problems with this specification, particularly for marketing and micro-econometric applications. In marketing applications, the θ vector will typically contain coefficients which measure the responsiveness of the cross-sectional unit to a marketing mix variable (such as price) as well as various intercepts for particular brands or products. One can well imagine that the typical distribution of a price response or sensitivity coefficient would be highly skewed and with support only over non-positive values.[3] Clearly, the normal distribution would not be well-suited to approximate the distribution of price coefficients. Similarly, the normal distribution is not ideally suited to approximate the distribution of brand intercepts across consumers, as the distribution of brand intercepts is apt to be multi-modal. That is, we might expect there to be groups of consumers who prefer a particular brand and other groups who do not exhibit a high degree of brand loyalty.

The basic assumption behind a random coefficient model is that the coefficients for each cross-sectional unit are drawn independently from one common hypothesized distribution.

---

[3]The log-normal distribution can be used for any coefficient via the reparameterization, $\theta = e^{\delta}$; however, a dependent mixture of log-normal and normal distributions for the entire θ vector will require non-conjugate prior formulations.

This assumption has been widely criticized as unrealistic. There may be characteristics of the cross-sectional unit which alter the random coefficient distribution. If these same characteristics also affect the distribution of the x variables, then methods which impose independence between the random coefficients and the independent variables can be both inefficient (they fail to exploit the information in the levels of the x variables which should be useful to make inferences about the random coefficient) and inconsistent. For example, suppose that the model under consideration is a basic demand or sales response model in which the dependent variable represents quantity demanded by each consumer and the critical independent variable is price. Suppose consumer wealth is unobserved. If we think that wealth is related to both the prices that consumer is exposed to as well as to the responsiveness of the consumer to price, then our standard random coefficient model (even with a very flexible random coefficient distribution) is mis-specified. Those consumers who are wealthy, for example, may have smaller price response coefficients and be exposed to higher prices than those consumers who are relatively poor. This yields a standard omitted-variable bias in the estimation of the price coefficient which, in this example, would be positive and yield an inconsistent level of the random coefficient distribution for the price effect.

This basic weakness of the standard random coefficient model has lead to widespread use of so-called "fixed" effects models. That is, rather than impose any specific random coefficient model, each cross-sectional unit is simply analyzed independently, thereby avoiding the assumptions of the random coefficient model. The fixed effects approach is not practical for panels with a short time dimension as the individual independent model estimates will be subject to large sampling errors; in some cases, the MLE may not even exist at the panelist level if there is insufficient variation in the independent variables or if the full range of y is not observed. The practice of confining heterogeneity or variability in θ only to the intercept is only useful for linear models in which differencing can be used to remove the intercepts from estimation. In a more general non-linear model, there are no analogues of differencing and there is no real advantage to allowing only the intercepts to vary across panel units.

The methods for approximating multivariate densities developed in Chapters 1 and 2 can be applied to random coefficient models to free these models from overly restrictive assumptions regarding the random coefficient distribution. This removes one major objection to the random coefficient approach but does not remove objections to the assumption of independence between the regressors and the random coefficient distribution. If there are observable characteristics of the panel unit, it is a simple matter to incorporate them into the random coefficient distribution.

$$\theta_j = \Delta' z_j + u_j \qquad\qquad (5.1.4)$$

$$u_j \sim \text{Mixture of Normals} \qquad\qquad (5.1.5)$$

This specifies a multivariate regression relationship between the θ vector and observable characteristics, z_j with residuals that are distributed as a mixture of normals (either finite or infinite mixture). Posterior inference for Δ can easily be accomplished conditional on the indicator vector for the normal mixture in a standard Gibbs sampler as laid out in section 4.5 of Rossi, Allenby, and McCulloch (2005). This removes some of the force of the independence objection to the use of random coefficient models.

In the econometrics literature on linear models, there has been some emphasis on the comparison of fixed effects models with normal random coefficient models, in particular, as the basis for the Hausman specification test (see Hausman (1978) and Hausman and Taylor (1981)). That is, as a diagnostic, a researcher should compare the fixed effects estimates with those obtained from a normal random effects model. In this literature, there are typically only random intercepts with the slope coefficients viewed as common across cross-sectional units. Estimates of these common coefficients are compared in the

Hausman specification test. However, there is ample evidence that models with common and non-random coefficients are mis-specified and at variance with the typical empirical finding that all model coefficients vary across units. It is certainly more reasonable to interpret common coefficients as the mean of the random effect distribution. With more modern tools, the generalization of the Hausman approach would be to compare estimates of the common coefficients in a standard differencing approach to a linear model with the mean of the random effects distribution fit to the same data. The `bayesm` routine, `rhierLinearMixture`, implements a finite mixture of normals model for a panel data structure of linear regressions with the possibility of observables which change the mean of the coefficient distribution.

5.2 Semi-parametric Random Coefficient Logit Models

Consumer- or customer-level demand data is discrete, deriving from the fact that on any given purchase occasion consumers only have positive demand for a small subset of available products. If the set of products is relatively homogeneous or represent a highly substitutable set of products, then it may be reasonable to model demand as a mutually exclusive choice model with a multinomial distribution of product demand given product characteristics. The most popular model in this literature is the Multinomial Logit (MNL) model. An extensive literature on random coefficient logit models has been developed. This literature finds that there is considerable heterogeneity in demand parameters across consumers. That is, patterns of product loyalty and response to marketing mix variables such as price and advertising vary dramatically over consumers. In most of this literature, a normal model for heterogeneity has been assumed.[4]

---

[4]Kamakura and Russell (1989) use a discrete distribution to approximate the distribution of coefficients across consumers. Heckman and Singer (1984)

In this section, I will outline the semi-parametric Bayesian approach to a random coefficient logit model.

The semi-parametric random coefficient logit model starts with an MNL model conditional on cross-sectional unit parameters, θ^b.

$$\Pr(y_{jt} = l) = \frac{\exp\left(x'_{ljt}\theta^b\right)}{\sum_k \exp\left(x'_{kjt}\theta^b\right)} \qquad (5.2.1)$$

Here there are $k = 1, \ldots, K$ choice possibilities for the MNL dependent variable, y_{jt}, each product is indexed by j and each purchase occasion by t. A semi-parametric approach is based on the recognition that the $\{\theta^b\}$ are sufficient for any model of the random coefficient distribution. If a finite or infinite mixture of normals is used, then a standard data augmentation with the indicator of the normal component can be used. Conditional on this indicator, there is a specific normal prior for each cross-sectional unit b. That is, the mixture of normals approach specifies a conditional normal prior for each cross-sectional unit with groups of units clustered with the same normal prior. In the MCMC algorithm, the indicator variable (which specifies which prior is used for each cross-sectional unit) is also drawn

---

provide a semi-parametric interpretation of this idea and suggest that if the number of mass points in the discrete approximation to the distribution coefficients is allowed to increase with the sample size that the discrete model can "approximate" any distribution. Given that the approximation is discrete and the actual random coefficient distribution is continuous, the Heckman and Singer results only apply to the convergence of certain functionals of the random coefficient distribution. In addition, the Heckman and Singer approach requires a rule which links the number of points in the discrete approximation to the sample size. As a practical matter, as Allenby and Rossi (1999) point out, there are many limitations of a discrete approximation approach. Many mass points will be required for high dimensional random coefficient distributions. One can think of the DP infinite mixture of normals as a generalization of the Heckman and Singer perspective, except where the random coefficient distribution is approximated with a continuous approximation and there is no need for ad hoc rules to expand the number of components in the mixture as the sample size increases.

providing the appropriate inference in the face of uncertainty with respect to the classification of each cross-sectional unit. In other words, given the draws of $\{\theta^h\}$, everything proceeds as outlined in Chapters 1 and 2 (see section 4.5 of Rossi, Allenby, and McCulloch (2005) for details). The only difficulty occurs with respect to the draws of the cross-sectional unit logit parameters.

The draws of θ^h can be accomplished by a Metropolis-Hastings random walk method. The only limitation is that the M-H random walk methods only work well if the random walk increments can be tuned to conform as closely as possible to the curvature in the conditional posterior (here y_h is the vector of all observations on the dependent variable for the hth cross-sectional unit).

$$p\,(\theta_h|y_h,\tau) \propto p\,(y_h|\theta_h)\,p\,(\theta_h|\tau) \qquad (5.2.2)$$

Therefore, for all except the most regular models, it will be necessary to customize the Metropolis chains for each cross-sectional unit. Without prior information on highly probable values of the first stage prior parameters, τ, it will be difficult to use the strategy of trial runs to tune the Metropolis chains given that a large fraction of cross-sectionals have limited information about the model parameters. One other possibility that is often employed is to use the pooled likelihood for all units and scale the Hessian from this pooled likelihood for the number of observations in any one unit (see Allenby and Rossi (1993)). Define $\bar{\ell}(\theta) = \prod_{h=1}^{H} \ell\,(\theta_h|y_h)$ as the pooled likelihood. The scaled Hessian is given by

$$\bar{H}_h = \frac{n_h}{N}\frac{\partial^2 log\,\bar{\ell}}{\partial\theta_h\partial\theta_h'}|_{\theta=\hat{\theta}_{MLE}}. \qquad (5.2.3)$$

$N = \sum_{h=1}^{H} n_h$. n_h is the number of observations for cross-sectional unit h. The virtue of the use of a Hessian based on the pooled sample is that the pooled MLE is often easy to find and has a non-singular Hessian.

The scaled Hessian is based on the the curvature of the pooled data likelihood function. This is influenced not only by the curvature of individual unit likelihoods but also by the difference between units. While the scaled Hessian in (5.2.3) will get the scaling or units approximately correct for each element of θ, there is no guarantee that this curvature estimate will approximate the correlation structure in each individual unit.

At the opposite extreme from the use of the pooled MLE would be to use Hessian estimates constructed from each unit likelihood. This would require that the MLE exist and that the Hessian is non-singular for each cross-sectional unit likelihood. For choice model applications, this would require, at a minimum, that each cross-sectional unit be observed to choose at least once from all choice alternatives (sometimes termed a "complete" purchase history). If a unit does not choose a particular alternative and if an alternative-specific intercept is included in the model, then the MLE will not be defined for this unit. There would exist a direction of recession in which an intercept will drift off to $-\infty$ with an increasing likelihood. What is required is a regularization of the unit-level likelihood for that sub-sample of units with singular Hessians or non-existent MLEs. Our proposal is to borrow from the "fractional" likelihood literature for the purpose of computing an estimate of the unit-level Hessian. This is only used for the Random Walk (RW) Metropolis increment covariance matrix and is *not* used to replace the unit-level likelihood in posterior computations.

To compute the Hessian, we form a fractional combination of the unit-level likelihood and the pooled likelihood.

$$\ell_h^*\left(\theta\right) = \ell_h\left(\theta\right)^{(1-w)}\bar{\ell}\left(\theta\right)^{w\beta} \tag{5.2.4}$$

The fraction, w, should be chosen to be a rather small number so that only a "fraction" of the pooled likelihood, $\bar{\ell}$, is combined with the unit likelihood, ℓ_h, to form the regularized likelihood. β is chosen to properly scale the pooled likelihood to the same order as the unit likelihood. $\beta = \frac{n_h}{N}$. (5.2.4) is maximized to estimate the Hessian at the "modified" MLE. This Hessian can be combined with the normal covariance matrix from the unit-level conditionally normal prior (note: if the prior is of the mixture of normal form, we are conditioning on the indicator for this unit). If the RW Metropolis increments are $N\left(0, s^2\Omega\right)$, then

$$\Omega = \left(H_h + V_\theta^{-1}\right)^{-1}, \tag{5.2.5}$$

$$H_h = -\frac{\partial^2 log\, \ell_h^*}{\partial\theta\,\partial\theta'}|_{\theta=\hat{\theta}_h}. \tag{5.2.6}$$

$\hat{\theta}_h$ is the maximum of the modified likelihood in (5.2.4).

The customized M-H method outlined above has been implemented along with the MNL logit likelihood in the `bayesm` routine, `rhierMnlRwMixture`, for a finite mixture of normals prior and in the routine, `rhierMnlDP`, for the infinite mixture of normals approach.

5.3 An Empirical Example of a Semi-parametric Random Coefficient Logit Model

Empirical researchers in marketing have documented a form of state dependence whereby consumers become "loyal" to products they have consumed in the past (see, for example, Guadagni and Little (1983); Seetharaman (2004); Seetharaman, Ainslie, and Chintagunta (1999); Erdem (1996)).[5] That is, consumers behave as though there is a utility premium from continuing to purchase the same product as they have purchased in the past or, equivalently, there is a psychological cost to switching products. However, it has not been established that this form of state dependence can be identified in the presence of consumer heterogeneity of an unknown form. Heckman (1981) pointed out that there can be a confounding of state dependence and

---

[5] This section is adapted from portions of Dubé, Hitsch, and Rossi (2010).

heterogeneity in parametric models of heterogeneity. That is, a mis-specified model of heterogeneity (i.e., a mis-specified random coefficient distribution) can lead to findings of spurious state dependence. Fundamentally, it is hard to tell the difference between brand loyalty (that is, high utility for some products) and state dependence. Are consumers "loyal" or simply state dependent?

The conditions for the separate identification of state dependence and heterogeneity require not only a panel data structure but also a source of exogenous shifts in the "state" or product consumed on the last purchase occasion. Ideally, we would like to experimentally manipulate the last purchase choice in order to trace out a state dependence effect and distinguish this from simple loyalty to the product. In many marketing panel data sets, there are frequent price promotions or discounts in which prices change radically. These price promotions are conducted at the store level and, by definition, cannot be related to consumer preferences. Therefore, these price changes can induce exogenous changes in product purchase akin to an experiment in which the experimenter was able to force consumers to switch away from favored brands.

The empirical literature on state dependence assumes a normal distribution of heterogeneity.[6] There is no particular reason to assume that distributions of taste parameters should exhibit symmetric and unimodal distributions. It might be argued, for example, that the distribution of brand intercepts should be multi-modal, corresponding to different relative brand preferences for different groups of consumers. In order to establish that state dependence findings are robust to distributional assumptions, we implement a very flexible, semi-parametric specification consisting of a mixture of multivariate normal distributions. While we argue that our Bayesian methods provide extreme flexibility while retaining desirable smoothness

---

[6]See, for example, Keane (1997), Seetharaman, Ainslie, and Chintagunta (1999), and Osborne (2007). Shum (2004) uses discrete distribution of heterogeneity.

properties, we also consider a form of model-free evidence that our heterogeneity specification is adequate.

5.3.1 Model and Econometric Specification

Our baseline model consists of households making discrete choices among J products in a category and an outside option each time they go to the supermarket. The timing and incidence of trips to the supermarket, indexed by t, are assumed to be exogenous. To capture inertia in choices, we take the standard approach, often termed "state dependent demand," and assume that current utilities are affected by the previous product chosen in the category. For ease of exposition, we drop the household-specific index below. In the empirical specification, all the model parameters will be household specific.

The utility index from product j at time period t is

$$u_{jt} = \alpha_j + \eta p_{jt} + \gamma I \{s_t = j\} + \epsilon_{jt} \qquad (5.3.1)$$

where p_{jt} is the product price[7] and ϵ_{jt} is the standard iid error term used in most choice models. In the model given by (5.3.1), the brand intercepts represent a persistent form of vertical product differentiation that captures a household's intrinsic product (or brand) preferences. $s_t \in \{1, \ldots, J\}$ summarizes the history of past purchases from the perspective of impact on current utility. If a household buys product k in period $t - 1$, then $s_t = k$. If the household chooses the outside option, then the household's state remains unchanged: $s_t = s_{t-1}$. Some term s_t the "loyalty" state of the household. If $s_t = j$, the household is said to be "loyal" to brand j. While the use of the last purchase as the summary of the past purchases is very frequently used in

---

[7]Other characteristics of the store environment facing the household could be entered into the "utility" model. But, then it may be problematic to interpret this as a utility specification. For example, many researchers include in-store advertising variables directly in the choice model.

empirical work, it is by no means the only possible specification. For example, Seetharaman (2004) considers various distributed lags of past purchases.

If $\gamma > 0$, then the model in (5.3.1) will generate a form of inertia. If a household switches to brand k, the probability of a repeat purchase of brand k is higher than prior to this purchase. One possible interpretation is that γ results from a form of psychological switching costs (see Farrell and Klemperer (2006)).

5.3.2 *Econometric Specification*

At the household level (indexed by h), we specify a multinomial logit model with the outside good expected utility set to zero.

$$\Pr(j) = \frac{\exp\left(\alpha_j^h + \eta_j^h Price_j + \gamma^h I\left\{s = j\right\}\right)}{1 + \sum_{k=1}^{J} \exp\left(\alpha_k^h + \eta_k^h Price_j + \gamma^h I\left\{s = k\right\}\right)} \quad (5.3.2)$$

If we denote the vector of household parameters $\left(\alpha_1^h, \ldots, \alpha_J^h, \eta^h, \gamma^h\right)$ by θ^h, then heterogeneity of household types can be accommodated by assuming that the collection of $\left\{\theta^h\right\}$ are drawn from a common distribution. In the empirical literature on state dependent demand, a normal distribution is often assumed, $\theta^h \sim N(\bar{\theta}, V_\theta)$. Frequently, further restrictions are placed on V_θ such as a diagonal structure (see, for example, Osborne (2007)). Other authors restrict the heterogeneity to only a subset of the θ vector. The use of restricted normal models is due, in part, to the limitations of existing methods for estimation of random coefficient logit models.

If normal models for heterogeneity are unable to capture the full distribution of heterogeneity, then there is the potential to create a spurious finding of inertia or the importance of state dependence. For example, consider the situation in which there is a bimodal distribution of preferences for a particular brand. One mode corresponds to a sub-population of consumers who find the brand relatively superior to other brands in the choice set, while the other mode corresponds to consumers who find this brand relatively inferior. The normal approximation to a bimodal distribution would be symmetric and centered at zero. The normal would understate the differences in brand preferences relative to the bimodal distribution. When applied to data, the model with the normal distribution would have a likelihood that puts mass on positive inertia parameter values in order to accommodate the observation that some households persistently buy (do not buy) one of the brands.

Rather than restricting the distribution of parameters across households, we want to allow for the possibility of non-normal and flexible distributions. This poses a challenging econometric problem. Even if we were to observe the $\{\theta^h\}$ without error, we would be faced with the problem of estimating a high dimensional distribution (in the applications below, we estimate models with 5–10 dimensional distributions). In practice, we have only imperfect information regarding household level parameters which adds to the econometric challenge. Even with hundreds of households, we may only have limited information for any one household given that there are typically not more than 50 observations per household. This requires a method that does not over-fit the data. One approach to the problem of over-fitting is to use proper prior distributions which create forms of smoothing and parameter shrinkage. Our approach is to specify a hierarchical prior with a mixture of normals as the first stage prior. The hierarchical prior provides one convenient way of specifying an informative prior which avoids the problem of over-fitting even with a large number of normal components. The first stage is a mixture of K multivariate normals and the second stage consists of priors on the parameters of the mixture of normals.

$$ p\left(\theta^h | \pi, \{\mu_k, \Sigma_k\}\right) = \sum_{k=1}^{K} \pi_k \phi\left(\theta^h | \mu_k, \Sigma_k\right) \qquad (5.3.3) $$

$$ \pi, \{\mu_k, \Sigma_k\} | b \qquad (5.3.4) $$

Here the notation ·|· indicates a conditional distribution and *b* represents the hyper-parameters of the priors on the mixing probabilities and the parameters governing each mixture component.

As is well-known, a mixture of normals models is very flexible and can accommodate deviations from normality such as thick tails, skewness, and multi-modality. A priori, we might expect that brand preference parameters (intercepts) might have a multi-modal distribution. The modes might correspond to subgroups of consumers who very much like, very much dislike, or who are indifferent to the brand. In addition, we might expect the distribution of price coefficients to be skewed to the left since consumers should behave in accordance with a negative price coefficient and there may be some extremely price sensitive consumers. At the same time, we do not expect preference parameters to be independent. Thus, we are faced with the task of fitting a multivariate mixture of normals.

Some might argue that you do not have a truly non-parametric method unless you can claim that your procedure consistently recovers the true density of parameters in the population of all possible households. In the mixture of normals model, this requires that the number of mixture components (K) increases with the sample size. Our approach is to fit models with successively larger numbers of components and gauge the adequacy of the number of components by examining the fitted density as well as the Bayes Factor (see model selection discussion below) associated with each number of components. What is important to note is that our improved MCMC algorithm is capable of fitting models with a large number of components at relatively low computational cost.

5.3.3 Posterior Model Probabilities

In order to establish that the inertia we observe in the data can be interpreted as a true state dependent utility, we will compare a variety of different specifications. Most of the specifications we will consider will be heterogeneous in that a prior distribution or random coefficient specification will be assumed for all utility parameters. This poses a problem in model comparison as we are comparing different and heterogeneous models. As a simple example, consider a model with and without the lagged choice term. This is not simply a hypothesis about a given fixed dimensional parameter, $H_0 : \gamma = 0$, but a hypothesis about a set of household level parameters. The Bayesian solution to this problem is to compute posterior model probabilities and compare models on this basis. A posterior model probability is computed by integrating out the set of model parameters to form what is termed the marginal likelihood of the data. Consider the computation of the posterior probability of model M_i:

$$p\,(M_i|D) = \int p\,(D|\Theta, M_i)\, p\,(\Theta|M_i)\, d\Theta \times p\,(M_i) \qquad (5.3.5)$$

where D denotes the observed data, Θ represents the set of model parameters, $p\,(D|\Theta, M_1)$ is the likelihood of the data for M_1, and $p\,(M_i)$ is the prior probability of model i. The first term in (5.3.5) is the marginal likelihood for M_i.

$$p\,(D|M_i) = \int p\,(D|\Theta, M_i)\, p\,(\Theta|M_i)\, d\Theta \qquad (5.3.6)$$

The marginal likelihood can be computed by reusing the simulation draws for all model parameters that are generated by the MCMC algorithm using the method of Newton and Raftery (1994).

$$\hat{p}\,(D|M_i) = \left(\frac{1}{R} \sum_{r=1}^{R} \frac{1}{p\,(D|\Theta, M_i)} \right)^{-1} \qquad (5.3.7)$$

$p\,(D|\Theta, M_i)$ is the likelihood of the entire panel for model i. In order to minimize overflow problems, we report the log of the trimmed Newton–Raftery MCMC estimate of the marginal likelihood. Bayesian model comparison can be done on the basis of the marginal likelihood (assuming equal prior model probabilities).

Posterior model probabilities can be shown to have an automatic adjustment for the effective parameter dimension. That is, larger models do not automatically have higher marginal likelihood as the dimension of the problem is one aspect of the prior that always matters. While we do not use asymptotic approximations to the posterior model probabilities, the asymptotic approximation to the marginal likelihood illustrates the implicit penalty for larger models (see, for example, Rossi, McCulloch, and Allenby (1996)).

$$log(p(D|M_i)) \approx log(p(D|\hat{\Theta}_{MLE}, M_i)) - \frac{p_i}{2}log(n) \quad (5.3.8)$$

p_i is the effective parameter size for M_i and n is the sample size. Thus, a model with the same fit or likelihood value but a larger number of parameters will be "penalized" in marginal likelihood terms. Choosing models on the basis of marginal likelihood can be shown to be consistent in model selection in the sense that the true model will be selected with higher and higher probability as the sample size becomes infinite.

5.3.4 Data

For our empirical analysis, we estimate the logit demand model described above using household panel data containing all purchase behavior for the refrigerated orange juice and the 16-ounce tub margarine categories. The panel data were collected by ACNielsen for 2,100 households in a large midwestern city between 1993 and 1995. In each category, we focus only on those households that purchase a brand at least twice during our sample period. Hence we use 355 households to estimate orange juice, and 429 households to estimate margarine demand.

Table 5.1 lists the products considered in each category as well as the purchase incidence, product shares, and average prices.

Table 5.1. Description of the Data

Margarine

Product	Average price	% Trips
Promise	1.69	13.11
Parkay	1.63	4.98
Shedd's	1.07	12.66
ICBINB	1.55	23.51
no-purchase (% trips)	45.73	
# households	429	
# trips per household	18.25	
# purchases per household	9.90	

Refrigerated Orange Juice

Product	Average Price	% Trips
64oz MM	2.21	11.1
Premium 64oz MM	2.62	7.00
96oz MM	3.41	14.7
Premium 64oz TR	2.73	28.8
64oz TR	2.26	6.76
Premium 96oz TR	4.27	7.99
no-purchase (% trips)	23.75	
# households	355	
# trips per household	12.3	
# purchases per household	9.37	

We define the outside good in each category as follows. In the refrigerated orange juice category, we define the outside good as any fresh or canned juice product purchase other than the brands of orange juice considered. In the tub margarine category, we define the outside good as any spreadable product, i.e., jams, jellies, margarine, butter, peanut butter, etc. In Table 5.1, we see a no-purchase share of roughly 24% in refrigerated juice and 46% in tub margarine. We use this definition of the outside good to model only those shopping trips where purchases in the product category are considered.

In our econometric specification, we will be careful to control for heterogeneity as flexibly as possible to avoid confounding loyalty with unobserved heterogeneity. Even with these controls in place, it is still important to ask which patterns in our consumer shopping panel give rise to the identification of inertial or state dependent effects. The marginal purchase probability is considerably smaller than the re-purchase probability for all products considered. While this evidence is consistent with inertia, it could also be a reflection of heterogeneity in consumer tastes for brands. The identification of inertia in our context relies on the frequent temporary price changes typically observed in supermarket scanner data. If there is sufficient price variation, we will observe consumers switching away from their preferred products. The detection of state dependence relies on spells during which the consumer purchases these less-preferred alternatives on successive visits, even after prices return to their "typical" levels.

We use the orange juice category to illustrate the source of identification of inertia or state dependence in our data. First, we observe spells during which a household repeat-purchases the same product. Conditional on a purchase, we observe 1889 such repeat-purchases out of our total 3328 purchases in the category. Second, we observe numerous instances during which a spell is initiated by a discount price. We classify each product's weekly prices as either "regular" or "discount," where the latter implies a temporary price decrease of at least 5%. Focusing on non-favorite products, i.e., products that are not the most frequently purchased by a household, nearly 60% of the purchases are for products offering a temporary price discount. We compare the repeat-purchase rate for spells initiated by a price discount (i.e., a household repeat-buys a product that was on discount when they previously purchased it) to the marginal probability of a purchase in Table 5.2. For all brands of Minute Maid orange juice, the sample re-purchase probability conditional on a purchase initiated by a discount is .74, which exceeds the marginal purchase probability of .43. The same is true for Tropicana brand products with the conditional re-purchase probability of .83

Table 5.2. Re-purchase Rates

Margarine

Brand	Purchase frequency	Re-purchase frequency	Re-purchase frequency after discount
Promise	.24	.83	.85
Parkay	.09	.90	.86
Shedd's	.23	.81	.80
ICBINB	.43	.88	.88

Orange Juice

Brand	Purchase frequency	Re-purchase frequency	Re-purchase frequency after discount
Minute Maid	.43	.78	.74
Tropicana	.57	.86	.83

compared to the marginal purchase probability of .57. This is suggestive that observed high re-purchase rates are not simply the result of strong brand preferences but are caused by some sort of inertia.

Inertia or persistence in brand choices can be viewed as one possible source of dependence in choices over time even for the same consumer. Another frequently cited source of non-zero order purchase behavior is household inventory holdings (see, for example, Erdem, Imai, and Keane (2003)). If households have some sort of storage technology, then they may amass a household inventory either to reduce shopping costs (assuming there is a category-specific fixed cost of shopping) or to exploit a sale or price discount of short duration. It should be emphasized that household stockpiling has implications for the quantity of purchases as well as the timing of purchases. Our state dependence formulation suggests that the specific brand purchased on the last shopping trip should influence the current brand choice. A model of stockpiling simply suggests that as the time between purchases increases the hazard rate of purchase should increase. We neither use the quantity of purchase nor the timing of purchase incidence in our analysis. Finally, we should note that the possibilities for household inventory of the products (especially refrigerated orange juice) appear to be limited. In our data, over 80 percent of all purchases are for one unit of the product, suggesting that stockpiling is not pervasive.

5.3.5 Heterogeneity and State Dependence

It is well-known that state dependence and heterogeneity can be confounded (Heckman (1981)). We have argued that frequent price discounts or sales provide a source of brand switching that can identify inertia or state dependence in choices separately from heterogeneity in household preferences. However, it is an empirical question as to whether or not inertia is an important force in our data. With a normal distribution of heterogeneity, a number of authors have documented that positive state dependence or inertia is present in CPG panel data (see, for example, Seetharaman, Ainslie, and Chintagunta (1999)). Frank (1962) and Massy (1966) document state dependence at the panelist level using older diary data. However, there is still the possibility that these results confirming inertia are not robust to controls for heterogeneity using a flexible or non-parametric distribution of preferences. Our approach is to fit models with and without an inertia term and with and without various forms of heterogeneity. It is particularly convenient that our mixture of normals approach nests the normal model in the literature.

Table 5.3 provides log marginal likelihood results that facilitate assessment of the statistical importance of heterogeneity and inertia. All log marginal likelihoods are estimated using a Newton–Raftery style estimator that has been trimmed of the top and bottom 1 percent of likelihood values as is recommended in the literature. We compare models without heterogeneity to a normal model (a one component mixture) and to five and ten mixture component models.

As is often the case with consumer panel data (Allenby and Rossi (1999)), there is pronounced heterogeneity. In a

Table 5.3. Log Marginal Likelihood for State Dependence (SD) Specifications

Model	Margarine	Orange juice
Homogeneous model without SD	−10755	−7612
5 comp Normal without SD	−5575	−4528
5 comp Normal with lagged prices, no SD	−5517	−4389
Homogeneous model with SD	−8175	−6297
5 comp Normal with SD	−5501	−4434
5 comp, SD, randomized purchase sequence	−5581	−4503
5 comp, SD, interaction with discount	−5537	−4419

model with an inertia or state dependence term included, the introduction of normal heterogeneity improves the model fit dramatically. The log marginal likelihood improves by more 20 percent when normal heterogeneity is introduced. If two models have equal prior probability, the difference in log marginal likelihood is related to the ratio of posterior model probabilities.

$$log\left(\frac{p\,(M_1|D)}{p\,(M_2|D)}\right) = log\,(p\,(D|M_1)) - log\,(p\,(D|M_2)) \quad (5.3.9)$$

Introduction of normal heterogeneity improves the log marginal likelihood by more than 100 points, such that the ratio of posterior probabilities is more than $exp\,(100)$, providing overwhelming evidence in favor of a model with heterogeneity in both product categories.

The normal model of heterogeneity does not appear to be adequate for our data as the log marginal likelihood improves substantially (by at least 50 points) when a five component mixture model is used. For example, for margarine products in a model with an inertia term, moving from one to five normal components increases the log marginal likelihood from −5613 to −5550. Remember that the Bayesian approach automatically adjusts for effective parameter size (see section 5.3.3) and the increase in log marginal density observed in Table 5.3 represents a meaningful improvement in fit.

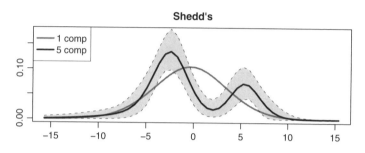

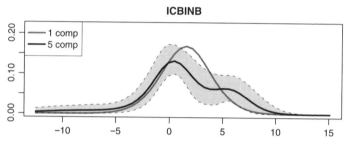

Figure 5.1. Margarine Intercepts

Figures 5.1–5.4 provide direct evidence on the importance of a flexible distribution of heterogeneity. Each figure plots the estimated marginal distribution of intercept, price, and inertia or "state dependence" coefficients from the five component mixture in black (here we use the posterior mean as the Bayes estimate of each density value). The envelope enclosing the five component marginal densities is a 90 percent point wise HPD region. The one component fitted density is drawn in medium grey. A number of the parameters exhibit a dramatic departure from normality. For example, the Shedd's brand of margarine has a noticeably bimodal marginal distribution across households. One mode is centered on a positive value (all intercepts should be interpreted as relative to the "outside" good which is defined as other products in the category) indicating strong brand preference for Shedd's. The other mode is centered

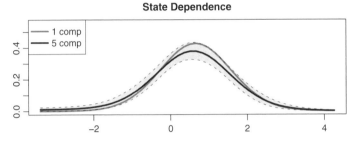

Figure 5.2. Price and State Dependence Coefficients: Margarine

on a value closer to zero, reflecting consumers who view Shedd's as comparable to other products in the category. One could argue that distributions with multiple modes are more likely to be the norm rather than the exception with any set of branded products. The price coefficient (Figures 5.2 and 5.4) is also non-normal, exhibiting pronounced left skewness. Again, it might be expected that there is a left tail of extremely price sensitive consumers. We note that the prior distribution for the price coefficient is symmetric and centered at zero.

Thus, there is good reason to doubt the appropriateness of the standard normal assumption for many choice model parameters. This opens the possibility that the findings documenting the importance of state dependence or inertia in choices are influenced, at least in part, by arbitrary distributional assumptions. However, the importance of the inertia or "state dependence"

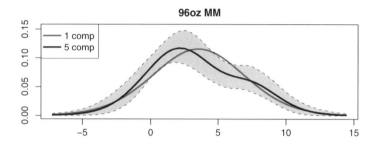

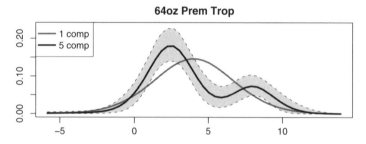

Figure 5.3. Refrigerated Orange Juice Brand Intercepts

remains even when a flexible five component normal is specified. The log marginal likelihood increases from -5575 to -5501 when inertia terms are added to a five component model for margarine and from -4528 to -4434 in refrigerated orange juice. Figures 5.2 and 5.4 show that the marginal distribution of the inertia parameter is well-approximated by a normal distribution for these two product categories. While this is not definitive evidence, it does suggest that the findings of inertia or state dependence in the literature are not artifacts of the normality assumption commonly used.

The five component normal mixture is a very flexible model for the joint density of choice model parameters. However, before we can make a more generic "semi-parametric" claim that our results are not dependent on the form of the distribution, we must provide evidence of the adequacy of the five component

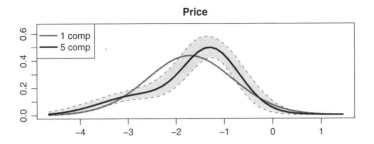

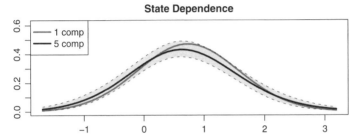

Figure 5.4. Price and State Dependence Coefficients for Refrigerated Orange Juice

distributional model. Our approach to this is to fit a ten component model. Many would consider this to be an absurdly highly parametrized model. For the margarine category, the ten component model would have a "nominal" number of 449 parameters (the coefficient vector is 8 dimensional).[8] The log marginal likelihood declines from five to ten components; from -5551 to -5559 for margarine and -4434 to -4435 for orange juice. These results marginally favor the 5 component model over the 10 component, but, more importantly, indicate no value from increasing the model flexibility beyond five components. We feel that the posterior model probability results in conjunction with

[8]There are $36 \times 10 = 360$ unique variance-covariance parameters plus 10×8 mean parameters plus 9 mixture probabilities $= 449$.

the high flexibility of the models under consideration justify the conclusion that we have accommodated heterogeneity of an unknown form.

5.3.6 Robustness Checks

State Dependence or a Mis-specified Distribution of Heterogeneity? Some may still doubt if we have indeed found inertia or if the lagged choice coefficient simply proxies for a mis-specification of the distribution of heterogeneity. We perform a simple check to test for this possibility. Suppose there is no state dependence and the coefficient on the lagged choice picks up taste differences across households that are not accounted for by the assumed functional form of heterogeneity. Then, if we randomly reshuffle the order of shopping trips, the coefficient on the lagged choice will not change and still provide misleading evidence for inertia. In Table 5.3 we show the log marginal likelihood for a five component model with an inertia term, which we fitted to our data with randomly reshuffled purchase sequences. The log marginal likelihood for the randomized sequence data is approximately the same as for the model without the inertia terms, and much lower than the log marginal density of the model with properly ordered data and the inertia term. We thus find strong evidence against the possibility that the lagged choice proxies for a mis-specified heterogeneity distribution.

State Dependence or Autocorrelation? While the randomized sequence test gives us confidence that we have found convincing evidence of a non-zero order choice process, it does not help distinguish between an inertia or state dependence model and a model with auto-correlated choice errors. Using normal models and a different estimation method, Keane (1997) finds that state dependent and auto-correlated error models produce very similar results. However, the economic implications of the two models are markedly different. With a structural interpretation for inertia as a form of state dependent utility, firms can influence the loyalty state of the customer and this has, for example, long-run pricing implications, while the auto-correlated errors model does not allow for interventions to induce inertia or loyalty to specific brands.

In order to distinguish between a model with a lagged choice or state dependent regressor and a model with auto-correlated errors, we implement the suggestion of Chamberlain (1985). We consider a model with a five component normal mixture for heterogeneity, no lagged choice or state dependent term, but including the lagged prices defined as the prices at the last purchase occasion. In a model with state dependence, price can influence the loyalty (or state) variable and this will influence subsequent choices. In contrast, in a model with auto-correlated errors, it is not possible to influence persistence in choices using exogenous variables. In Table 5.3, we compare the log marginal likelihood of the model without state dependence and a five component normal mixture with the log marginal likelihood of the same model including lagged prices. For margarine, the addition of lagged prices improves the log marginal likelihood by more than 50 points and by more than 100 points for refrigerated orange juice. This is strong evidence in favor of a "state dependence" specification with lagged choices.

A limitation of the Chamberlain suggestion (as noted by both Chamberlain himself and Erdem and Sun (2001)) is that consumer expectations regarding prices (and other right-hand side variables) might influence current choice decisions. Lagged prices might simply proxy for expectations even though there is no state dependence at all. Thus, the importance of lagged prices as measured by the log marginal likelihood is suggestive but not definitive.

As another comparison between a model with auto-correlated errors and a state dependent model, we exploit the price discounts or sales in our data. Since auto-correlated errors are not synchronized across households nor with price discounting by the retailer, we can differentiate between state dependent and auto-correlated error models by examining the impact of price discounts on measured state dependence. The intuition for this test is as follows. In a world of serially correlated errors, households that are induced by price discounts to switch to a new product will not exhibit inertia or persistence in choice. However, in a world with state dependence, brand switching induced by any reason should create persistence. To implement this idea, we interact the loyalty variable with an indicator for whether the loyalty state was initiated by a discount or not (i.e., whether the last product purchased was purchased on discount).

$$u_{jt} = \alpha_j + \eta_j p_{jt} + \gamma_1 I \{s_t = j\} + \gamma_2 I \{s_t = j\}$$
$$\cdot I \{discount_{s_t} = j\} + \epsilon_{jt} \qquad (5.3.10)$$

The term, $discount_{s_t}$, indicates whether the brand to which the consumer is currently loyal was on discount when it was last purchased. In a model with auto-correlated errors, the loyalty effect should dissipate for loyalty states generated by discounts, i.e., $\gamma_1 + \gamma_2 = 0$.

Table 5.3 provides a comparison of the log marginal likelihoods for the specification in (5.3.10); the log marginal likelihood values for the discount interaction term are in the last row of the table. The interaction term does improve model fit but by a modest 15–20 log density points. It remains an open question as to whether measured state dependence changes materially when we compare the distribution of the state dependence conditional on a past purchase that was or was not on discount. Recall that we allow for an entire distribution of parameters across the population of consumers so that we cannot provide the Bayesian analogue of a point estimate and a confidence interval. Instead, we plot the fitted marginal distribution of γ_1 and $\gamma_1 + \gamma_2$ in Figure 5.5. The black density curve is from our baseline model without any interaction term (5.3.1), the medium grey is the inertial or state dependence effect conditional on a discount on the focal brand during the previous purchase occasion (labelled "lagged sale") (denoted γ_1 in 5.3.10), and the dotted line is the effect conditional on no discount (labelled "no lagged sale") (denoted $\gamma_1 + \gamma_2$ in 5.3.10). There is little difference between the three densities for the orange juice category and a slight shift toward zero with a lagged sale in the

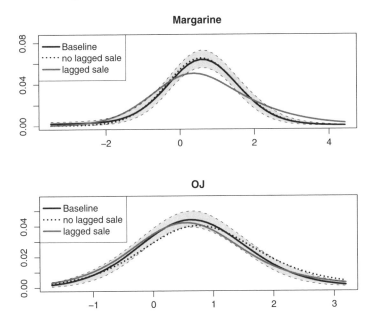

Figure 5.5. State Dependence and the Interaction with Price Discounts

margarine category. We conclude that there is scant evidence to support the claim that auto-correlated errors are the source of measured inertia.

Brand-Specific State Dependence

In the basic utility specification (5.3.1), the inertia effects are governed by one parameter that is constrained to be the same across brands for the same household. There is no particular reason to impose this constraint other than parsimony. Several authors have found the measurement of inertial effects to be difficult (see, for example, Keane (1997), Seetharaman, Ainslie, and Chintagunta (1999), and Erdem and Sun (2001)) even with a one component normal model for heterogeneity. The reason for imposing one "state dependence" or inertia parameter could simply be a need for greater efficiency in estimation. However, it would be misleading to report state dependent effects if these are limited to, for example, only one brand in a set of products. It also might be expected that some brands with unique packaging or trademarks might display greater inertia than others. It is also possible that the formulation of some products may induce more inertia via some mild form of "addiction" in that some tastes are more habit-forming than others. For these reasons, we consider an alternative formulation of the state dependence model with brand-specific loyalty parameters. Our Bayesian methods have a natural advantage for more highly parameterized models in the sense that if a model is weakly identified from the data, the prior keeps the posterior well-defined and regular.

A five component mixture of normals with brand specific inertia fits the data with a higher log marginal likelihood for both categories. For margarine, the log marginal likelihood increases from −5551 to −5505 when brand-specific effects are introduced into state dependence. There is an even more dramatic increase for the orange juice products, from −4436 to −4364. However, there is a difference between substantive and statistical significance. For this reason, we plot the fitted marginal densities for the inertia or "state dependence" parameters for each brand in Figures 5.6 and 5.7. The distributions displayed in Figure 5.6 compare the baseline model with models that allow for different inertia distributions for each brand. Interestingly, all four distributions are centered close to the baseline, constrained specification. In the orange juice category, Figure 5.7 plots the distributions of inertia parameters for the four highest share brands. In this category, the 96-ounce brands have higher inertia than the 64-ounce brands. We should note that the prior distribution[9] on the inertia parameters is centered

---

[9]It should be noted that our "prior" is a prior on the parameters of the mixture of normals—the mixing probabilities and each component mean vector and covariance matrix. This induces a prior on the distribution over parameters and the resultant marginal densities. While this is of no known analytic form, the fact that our priors on each component parameters are diffuse mean that the prior on the distributions is also diffuse.

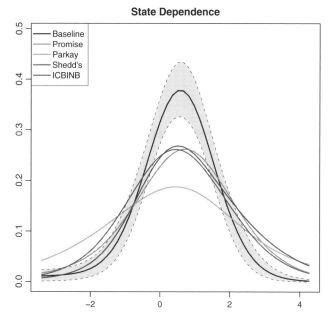

Figure 5.6. Brand-Specific State Dependence: Margarine

at zero and very diffuse. This means that data has moved us to a posterior which is much tighter than the prior and moved the center of mass away from zero. Thus, our results are not simply due to the prior specification but are the result of evidence in our data.

The main conclusion is that allowing for brand-specific inertia does not reduce the importance of inertial effects nor restrict these effects to a small subset of brands.

5.3.7 *Conclusions from Empirical Example*

Inertia in consumer purchases has been documented in a variety of studies that use frequently purchased consumer packaged

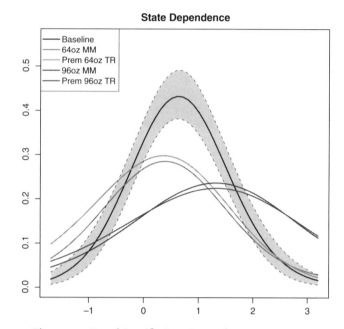

Figure 5.7. Brand-Specific State Dependence: Orange Juice

goods. Typically, the lagged brand choice is found to influence current brand choice positively and the results are interpreted as evidence in favor of state dependent demand. It is well-known, however, that state dependence can be confounded with heterogeneity. Households that simply prefer one brand over the others in a product category exhibit, what appears to be, persistence. In package goods panel data, there are frequent sales which can induce brand switching and which can, in principle, differentiate state dependence from heterogeneity.

Most empirical studies of state dependence use a normal distribution of heterogeneity. There is good reason to believe that a normal distribution may be inadequate to capture heterogeneity in many choice model parameters. It remains to be seen if the findings of state dependence are robust to a more flexible heterogeneity distribution. We use a mixture of a large number of multivariate normal distributions and exploit recent innovations in estimation technology to implement a Bayesian MCMC procedure for this problem.

When applied to data on choices among brands of tub margarine and refrigerated orange juice, we do, indeed, find very substantial evidence of non-normality. Findings of inertia in brand choice are robust to a flexible distribution of preferences. It might be argued, however, that our findings of inertia stem from auto-correlated choice error terms. As Keane (1997) has pointed out, "state dependence" and auto-correlated error specifications can produce similar patterns in the data. The key difference between state dependence and autocorrelation is that state dependence can be changed by external variables that alter the state and, therefore, the pattern of persistence in the future. On the other hand, the persistence stemming from auto-correlated errors cannot be altered. We exploit this difference between the models to create a test for autocorrelation based on whether the past purchase was initiated by a price discount or not. We find evidence in favor of the state dependent model and against the auto-correlated error specification.

The structural interpretation of the state dependence model is that, when a brand is purchased, there is a utility premium accorded that brand in future choices. Equivalently, we could interpret the state dependence model as a model of switching costs in which a cost is paid (in utility terms) from switching to brands not bought on the last purchase occasion. The switching cost interpretation of brand inertia or brand loyalty is based on the existence of psychological switching costs rather than explicit monetary or product adoption costs.

Alternative structural models which could give rise to inertia in choices are models with important brand search or learning effects. In search models, consumers may persist in purchasing one brand if the costs of exploring other options are high. In learning models, what appears to be inertia can arise because of imperfect information about product quality. Products which a consumer has consumed have less uncertainty in quality evaluation and this may make consumers reluctant to switch to alternative products for which there is greater quality uncertainty. In Dubé, Hitsch, and Rossi (2010), we show that the evidence in the data does not point to learning or search as the primary source of state dependence.

We have established a firmer basis for the structural interpretation of the state dependent choice model for demand. This model implies that variables under firm control, such as prices, can influence the future choice behavior of consumers. This opens a number of possibilities for work on firm policy. In the companion pieces, Dubé, Hitsch, Rossi, and Vitorino (2008) and Dubé, Hitsch, and Rossi (2009), we explore the implications of the estimates switching costs for dynamic pricing under multi-product monopoly and dynamic oligopoly, respectively.

6
Conclusions and Directions for Future Research

The preceding chapters establish two important conclusions: (1) mixture of normal models with appropriately assessed informative priors provide a very useful density approximation method that is applicable in high dimensions without undue over-fitting; and (2) many important problems in micro-econometrics can be solved by using mixture of normals to approximate key densities in these models. The applications considered so far include non-parametric regression with a continuous dependent variable, semi-parametric regression with a single index model, semi-parametric inference for panel choice models with unknown distributions of heterogeneity, and semi-parametric inference for instrumental variable models with linear structural and instrument equations.

6.1 When Are Non-parametric and Semi-parametric Methods Most Useful?

Non-parametric and semi-parametric work is demanding of data in the sense that there must be not only a large number of observations but, also, that non-parametric methods, however potent, cannot make precise inferences in regions where the data are sparsely distributed. To review the problem of sparseness, recall the non-parametric approach to regression outlined in Chapter 3. In the "full-information" approach I endorse, we model the joint distribution of the dependent and independent variables using flexible finite or infinite normal mixtures and then compute the posterior distribution of the regression function implied by the approximated joint distribution. This provides a fully flexible or non-parametric model in the sense that the conditional distribution implied by the fitted joint density can have any shape and the regression function (the conditional mean function) is not restricted to be of any particular parametric form. The shape of the regression function is informed "locally" by the joint density of the data around any particular point in the space. That is, $\mathbb{E}[y|x = x^*]$ is determined by the density of the independent variable data in the vicinity of x^*. If there is little data in the neighborhood of x^*, then the non-parametric method will attempt to interpolate. Obviously, the interpolation is problematic for a fully non-parametric method. The posterior distribution of the regression function value will properly reflect uncertainty and show a much wider distribution for conditional mean values in areas with sparseness as opposed to regions of densely packed x data. While it is comforting that the Bayesian methods advocated here will properly reflect uncertainty that stems from both small samples and sparseness, this does not eliminate the problems associated with applying non-parametric methods to econometric models.

For this reason, I have been selective in the sorts of problem for which I have advocated the use of semi-parametric and non-parametric methods. Consider the problem of modeling panel data on choices. At least three sorts of parametric assumptions are used in modeling choice data: (1) for a given panelist, we often assume that the multinomial logit model is appropriate to model the multinomial distribution of choices; (2) the multinomial logit model uses a linear index of observed characteristics of the panelists and choice alternatives; and (3) the distribution of the logit model preference or choice parameters is assumed to be multivariate normal across panelists. I have chosen to remove the third assumption (the normal heterogeneity distribution) and illustrate how a much more flexible and realistic model can be built based on a mixture approach. The decision to attack the third assumption is not purely arbitrary. In both micro-econometric and marketing applications, we typically have "short" panels with a large number of panelists. The large number of panelists allow for the use of flexible models for the distribution of parameters across panelists without risking very imprecise inferences. In principle, one could relax the assumptions of a linear index model as well as the logit probability form. However, there is a relatively small amount of information for each panelist. While it is possible to use mixture models to relax the first two assumptions, I would be reluctant to apply such a highly flexible model to panel data since inferences at the panelist level are apt to be very imprecise.

To summarize, semi- and non-parametric methods are most fruitfully employed in situations with relatively large amounts of data. However, what constitutes a "large amount" of data is dependent on the ability to impose reasonable and self-calibrating smoothness priors. In my experience, standard kernel density methods are difficult to apply in more than a few dimensions and often require many thousands of observations. With easily assessed proper priors, the Bayesian methods here can work well in 5–15 dimensions with much more modest sample sizes of 500+. In micro-econometric and marketing applications, these situations are typically large cross-sectional data sets or for panel data where the approach is semi-parametric and uses the cross-sectional dimension of the data.

6.2 Semi-parametric or Non-parametric Methods?

When should non-parametric as opposed to semi-parametric methods be used? One legitimate, but naive, point of view is that non-parametric methods should be employed as these methods make fewer arbitrary assumptions. I characterize this point of view as naive in at least two respects. One, non-parametric methods do rely on smoothing assumptions (such as there exists a joint density of the data) and may require stronger smoothing priors. Second, investigators often do not have sufficient data to use a fully non-parametric approach. Given the natural tendency of all serious empirical researchers to specify ever more complicated models, it may also be naive to assume that large enough data sets will ever be available for a fully non-parametric approach. For these reasons, semi-parametric methods will always be important for applied research. In the past, it might have been argued that we lack sufficient data and powerful enough methods for practical application of either semi-parametric or non-parametric methods, but I hope that I have made a strong argument by demonstration that this is no longer the case. Today, the limitations are more data-based than procedural.

In a semi-parametric approach, parametric assumptions are made about some aspects of the model with non-parametric methods applied to only part of the model. All true Bayesian approaches are likelihood-based so this means that in a semi-parametric Bayesian model we must make explicit assumptions regarding all aspects of the model. In the Bayesian approach, the distinction between parametric and semi-parametric methods is that part of the model is represented by a flexible form which, in principle, frees this aspect of the model from any finite parameterization. For example, a semi-parametric approach to regression could assume a linear index model but with an additive error whose distribution is not specified, but rather, approximated by an infinite mixture of normals (see section 4.1). A slightly more general version of this model would be to specify an unknown bivariate distribution of $(y, x'\beta)$ and approximate this by an infinite mixture of normals. This relaxes the additive error distribution but imposes the restriction that any conditional heterogeneity in the distribution of $y|x$ must come as a function of the linear index, $x'\beta$.

The essence of the semi-parametric approach is to reduce the dimension of the problem for which truly infinite-dimensional, non-parametric methods are required. In the regression example discussed above, we have reduced the dimension of the problem from $k + 1$ dimensional (modeling the joint distribution of y, x) to two or one dimension, depending on whether or not an additive error is used. There are many situations for which a two-dimensional or one-dimensional density approximate is feasible.

If the investigator is willing to live with a linear approximation to the conditional mean, then there are many situations in which the semi-parametric approach to regression will be appropriate but where the fully non-parametric methods will be infeasible or result in imprecise inference.

Another useful way of seeing the distinction between semi-parametric and non-parametric models is to contrast conditional and unconditional models. As we have seen, the non-parametric approach to regression starts from the joint or unconditional distribution of y, x and regards this as the fundamental object to be approximated. This buys quite a bit of flexibility but might require the approximation of a high dimensional joint distribution. The linear index approach reduces the dimension of the problem by conditioning on the linear index as sufficient for x or restricting the distribution of y to depend on x only thru the linear index. I have argued that the approach of building arbitrary flexibility into the conditional distribution of $y|x$ creates a much more complicated problem with no simple non-parametric solution; that is, the conditional distribution must have a shape which is driven in an arbitrary and completely flexible way by changes in x. This is difficult to achieve, while the joint non-parametric approach is very straightforward. For this reason, I believe that conditional models should almost always be viewed as compatible with semi-parametric methods.

6.3 Extensions

In general, a mixture of normals (either finite or infinite) can be used for any unknown continuous joint distribution. I have provided examples involving random coefficients and error terms.[1]

---

[1] Any model with a normal error term can be extended to the semi-parametric case using normal mixtures. See Chib (2008) for many examples of normal-based parametric models for panel data. Virtually all of the models considered in Chib can be extended using mixtures of normal error terms.

It is easy to extend the ideas considered here to models with a latent continuous structure.

For example, consider the latent formulation of a binary choice model.

$$y_i^* = x_i'\beta + \varepsilon_i \tag{6.3.1}$$

$$y_i = \begin{cases} 1 \ if \ y_i^* > 0 \\ 0 \ if \ y_i^* \leq 0 \end{cases} \tag{6.3.2}$$

A binary logit or binary probit model uses specific parametric models for the conditional distribution, $y_i^*|x$. (Note that the logistic distribution can be well-approximated by a mixture of normals [see Fruhwirth-Schnatter (2010) for details] and the probit model is based on one component normal model.) It is a simple matter to replace these parametric models with a mixture of normals. Data augmentation can be used to augment the model in (6.3.2) with the latent continuous dependent variable and an indicator vector for the mixture component. It is well-known that the model in (6.3.2) is not identified. Typically, iden-tification is achieved by scale normalization (setting the variance of ε to fixed value). However, identification can also be achieved fixing one of the regression coefficients to a convenient normal-ized value such as 1 or -1. This requires prior information on the sign of one of the coefficients (such as the price coefficient which might be presumed to be negative). Using this normalization, the mixture of normals model can be directly applied to the distribution of the latent error term without modification.

The binary choice model can be extended to a multinomial setting by increasing the dimension of the latent regression dependent variable to p or $p-1$, where p is the number of choices. In the multinomial choice model, a multivariate mixture of normals could be used for the underlying latent re-gression error term. This would provide increased flexibility over the Multinomial Probit Model (MNP). However, even with only one normal component, the identification of the off-diagonal elements of the latent covariance matrix (or correlations between latent choice errors) is tenuous at best, requiring large data sets. I am skeptical of the value of moving to a mixture of normals approach as a way of generalizing the MNP model, for this reason.

The model in (6.3.2) could easily be elaborated to include the possibility of a right-hand side endogenous variable in the choice model using the same basic ideas as in section 4.2.

$$x_i = z_i'\delta + \varepsilon_{x,i} \tag{6.3.3}$$

$$y_i^* = \beta x_i + w_i'\gamma + \varepsilon_{y,i} \tag{6.3.4}$$

$$y_i = \begin{cases} 1 \ if \ y_i^* > 0 \\ 0 \ if \ y_i^* \leq 0 \end{cases} \tag{6.3.5}$$

Here I consider the case of only one endogenous right-hand side variable with the possibility of any number of instruments (represented by z). A finite or infinite mixture of normals can be used to flexibly approximate the joint distribution of the error terms, ε_x, ε_y. Identification can be achieved by fixing a coefficient in the latent "structural" equation. Given that only the sign of latent variable structural equation dependent variable (y^*) is observed, this model specification is likely to require a great deal of data (both in terms of the numbers of observations and variation in the "exogenous" variables, z, w) to distinguish any meaningful departures from normality in the shape characteristics of the joint distribution of instrument and structural equation errors. While there is a great of deal of interest in instrumental variable methods for discrete dependent variable models, it is not clear that semi-parametric methods will prove to provide strikingly different inferences than strictly parametric models, given the coarseness of the mapping from the underlying latent structure to the observed discrete dependent variables.

Finally, a semi-parametric approach can be used to free the sample selection models of Heckman (1979) from the assumption of joint normal distribution on the errors in a system of two equations.

$$z_i^* = w_i'\gamma + u_i \qquad\qquad (6.3.6)$$

$$y_i = x_i'\gamma + \varepsilon_i \qquad\qquad (6.3.7)$$

$$y_i \text{ observed } if\ z_i^* > 0 \qquad\qquad (6.3.8)$$

A great deal of attention has been given to relaxing the assumption that the two error terms are bivariate normal in this model. In particular, there has been an emphasis in the classical econometrics literature on methods that do not make any explicit assumptions about the joint distribution and do not attempt to explicitly model this distribution. Clearly, there is a loss of efficiency when non-likelihood methods are used. It is a simple matter to utilize a mixture of normals approach for this model and retain the efficiency of a full likelihood-based approach with a very flexible error distribution (see Hasselt (2011) for details).

Thus, it is clear that, in many important econometric contexts, a mixture of normals distributional approximation can be used instead of a multivariate normal distribution. With proper priors, the mixture of normals provides great flexibility with little loss of efficiency due to over-fitting. The important question then is not whether or not to apply semi-parametric or non-parametric methods, but, rather, when will the data be informative enough to show the more flexible semi-and non-parametric methods to advantage in the sense that the data will reveal departures from standard parametric assumptions? In my view, this contrasts sharply with the classical non-parametric and semi-parametric literature that either defaults to impractical kernel smoothing methods or attempts to find consistent but inefficient methods which do not exploit the full information content of the data.

Bibliography

ABE, M. (1995): "A Nonparametric Density Estimation Method for Brand Choice Using Scanner Data," *Marketing Science*, 14(3), 300–325.

ABRAMOWITZ, M., AND I. A. STEGUN (eds.) (1964): *Handbook of Mathematical Functions*. Dover Press, 9th ed.

ACEMOGLU, D., S. JOHNSON, AND J. A. ROBINSON (2001): "The Colonial Origins of Comparative Development: An Empirical Investigation," *American Economic Review*, 91, 1369–1401.

ALLENBY, G. M., AND P. E. ROSSI (1993): "A Bayesian Approach to Estimating Household Parameters," *Journal of Marketing Research*, 30(2), 171–182.

——— (1999): "Marketing Models of Consumer Heterogeneity," *Journal of Econometrics*, 89, 57–78.

ANDREWS, D. F., AND C. L. MALLOWS (1974): "Scale Mixtures of Normal Distributions," *Journal of the Royal Statistical Society, Series B*, 36(1), 99–102.

ANDREWS, D. W. K., M. J. MOREIRA, AND J. H. STOCK (2006): "Optimal Two-sided Invariant Similar Test for Instrumental Variables Regression," *Econometrica*, 74(3), 715–752.

ANTONIAK, C. E. (1974): "Mixtures of Dirichlet Processes with Applications to Bayesian Nonparametric Problems," *The Annals of Statistics*, 2(6), 1152–1174.

BEKKER, P. A. (1994): "Alternative Approximations to the Distributions of Instrumental Variables Estimators," *Econometrica*, 63, 657–681.

BERRY, S., J. LEVINSOHN, AND A. PAKES (1995): "Automobile Prices in Market Equilibrium," *Econometrica*, 63(4), 841–890.

BERRY, S. T. (1994): "Estimating Discrete-Choice Models of Product Differentiation," *Rand Journal of Economics*, 25(2), 242–262.

BLACKWELL, D., AND J. B. MACQUEEN (1973): "Ferguson Distribtutions Via Polya Urn Schemes," *The Annals of Statistics*, 1(2), 353–355.

CARD, D. (1995): "Using Geographic Variation in College Proximity to Estimate the Return to Schooling," in *Aspects of Labor Market Behavior*, ed. by N. Christofides and R. Swidinsky, pp. 201–222. University of Toronto Press.

CHAMBERLAIN, G. (1985): "Heterogeneity, Omitted Variable Bias, and Duration Dependence," in *Longitudinal Analysis of Labor Market Data*, ed. by J. J. Heckman and B. Singer, chap. 1, pp. 3–38. Cambridge University Press.

CHAO, J. C., AND P. PHILLIPS (1998): "Posterior Distributions in Limited Information Analysis of the Simultaneous Equations Model Using the Jeffreys Prior," *Journal of Econometrics*, 87, 49–86.

CHERNOZHUKOV, V., AND C. HANSEN (2008): "The Reduced Form: A Simple Approach to Inference with Weak Instruments," *Economics Letters*, 100, 68–71.

CHIB, S. (2008): "Panel Data Modelings and Inference: A Bayesian Primer," in *The Econometrics of Panel Data*, ed. by L. Mátyás and P. Severstre, chap. 15, pp. 479–515. Springer-Verlag.

CONLEY, T. G., C. B. HANSEN, R. E. MCCULLOCH, AND P. E. ROSSI (2008): "A Semi-Parametric Bayesian Approach to the Instrumental Variable Problem," *Journal of Econometrics*, 144, 276–305.

DANIELS, M. J., AND R. E. KASS (1999): "Nonconjugate Bayesian Estimation of Covariance Matrices and Its Use in Hierarchical Models," *Journal of the American Statistical Association*, 94(448), 1254–1263.

DIEBOLT, J., AND C. P. ROBERT (1994): "Estimation of Finite Mixture Distributions through Bayesian Sampling," *Journal of the Royal Statistical Society, Series B*, 56(2), 363–375.

DUBÉ, J.-P., G. HITSCH, AND P. E. ROSSI (2009): "Do Switching Costs Make Markets Less Competitive?," *Journal of Marketing Research*, 46(4), 435–445.

——— (2010): "State Dependence and Alternative Explanations for Consumer Inertia," *RAND Journal of Economics*, 41(3), 417–445.

DUBÉ, J.-P., G. HITSCH, P. E. ROSSI, AND M. A. VITORINO (2008): "Category Pricing with State-Dependent Utility," *Marketing Science*, 27(3), 417–429.

ERDEM, T. (1996): "A Dynamic Analysis of Market Structure Based on Panel Data," *Marketing Science*, 15(4), 359–378.

ERDEM, T., S. IMAI, AND M. P. KEANE (2003): "Brand and Quantity Choice Dynamics Under Price Uncertainty," *Quantitative Marketing and Economics*, 1, 5–64.

ERDEM, T., AND B. SUN (2001): "Testing for Choice Dynamics in Panel Data," *Journal of Business and Economic Statistics*, 19(2), 142–152.

ESCOBAR, M. D., AND M. WEST (1995): "Bayesian Density Estimation and Inference Using Mixtures," *Journal of the American Statistical Association*, 90(430), 577–588.

FARRELL, J., AND P. KLEMPERER (2006): "Coordination and Lock-In: Competition with Switching Costs and Network Effects," *draft, Handbook of Industrial Organization 3*.

FERGUSON, T. S. (1974): "A Bayesian Analysis of Some Nonparametric Problems," *The Annals of Statistics*, 1(2), 209–230.

FERNANDEZ, C., AND M. F. J. STEEL (2000): "Bayesian Regression Analysis with Scale Mixtures of Normals," *Econometric Theory*, 16, 80–101.

FRANK, R. E. (1962): "Brand Choice as a Probability Process," *The Journal of Business*, 35(1), 43–56.

FRUHWIRTH-SCHNATTER, S. (2001): "Markov Chain Monte Carlo Estimation of Classical and Dynamic Switching and Mixture Models," *Journal of the American Statistical Association*, 96(453), 194–209.

——— (2010): "Data Augmentation and MCMC for Binary and Multinomial Logit Models," in *Statistical Modeling and Regression Structures*, ed. by T. Kneib and G. Tutz, pp. 111–132. Springer-Verlag.

GELMAN, A., J. B. CARLIN, H. S. STERN, AND D. B. RUBIN (2004): *Bayesian Data Analysis*. Chapman and Hall.

GELMAN, A., AND X. MENG (1991): "A Note on Bivariate Distributions that are Conditionally Normal," *American Statistician*, 45(2), 125–126.

GEWEKE, J. (1996): "Bayesian Reduced Rank Regression in Econometrics," *Journal of Econometrics*, 75, 121–146.

GEWEKE, J. (2005): *Contemporary Bayesian Econometrics and Statistics*. John Wiley & Sons.

GEWEKE, J., AND M. KEANE (2007): "Smoothly Mixing Regressions," *Journal of Econometrics*, 138, 252–290.

GHOSH, J. K., AND R. V. RAMAMOORTHI (2003): *Bayesian Nonparametrics*, Springer Series in Statistics. Springer, New York.

GUADAGNI, P. M., AND J. D. C. LITTLE (1983): "A Logit Model of Brand Choice Calibrated on Scanner Data," *Marketing Science*, 2(3), 203–238.

HASSELT, M. V. (2011): "Bayesian Inference in a Sample Selection Model," *Journal of Econometrics*, 165, 221–232.

HAUSMAN, J. (1978): "Specification Tests in Econometrics," *Econometrica*, 46, 1251–1271.

HAUSMAN, J., AND W. TAYLOR (1981): "Panel Data and Unobservable Individual Effects," *Econometrica*, 49, 1377–1398.

HECKMAN, J. J. (1979): "Sample Selection Bias as Specification Error," *Econometrica*, 47(1), 153–162.

—— (1981): "The Incidental Parameters Problem and the Problem of Initial Conditions in Estimating a Discrete Time-Discrete Data Stochastic Process and Some Monte Carlo Evidence," in *Structural Analysis of Discrete Data with Econometric Applications*, ed. by C. Manski and D. McFadden, chap. 4, pp. 179–195. MIT Press.

HECKMAN, J. J., AND B. SINGER (1984): "A Method for Minimizing the Impact of Distributional Assumptions in Econometric Models," *Econometrica*, 52(2), 271–320.

HOOGERHEIDE, L., F. KLEIBERGEN, AND H. V. DIJK (2007): "Natural Conjugate Priors for the Instrumental Variables Model Applied to Angrist-Krueger Data," *Journal of Econometrics*, 138, 63–103.

KAMAKURA, W. A., AND G. J. RUSSELL (1989): "A Probabilistic Choice Model for Market Segmentation and Elasticity Structure," *Journal of Marketing Research*, 26(4), 379–390.

KEANE, M. P. (1997): "Modeling Heterogeneity and State Dependence in Consumer Choice Behavior," *Journal of Business and Economic Statistics*, 15(3), 310–327.

KLEIBERGEN, F. (2002): "Pivotal Statistics for Testing Structural Parameters in Instrumental Variables Regression," *Econometrica*, 70, 1781–1803.

—— (2007): "Generalizing Weak Instruments Robust IV Statisics towards Multiple Parameters, Unrestricted Covariance Matrices, and Identification Statistics," *Journal of Econometrics*, 139, 181–216.

KOOP, G. (2003): *Bayesian Econometrics*. John Wiley & Sons.

LI, M., AND J. L. TOBIAS (2011): "Bayesian Methods in Microeconometrics," in *Handbook of Bayesian Econometrics*, ed. by J. Geweke, G. Koop, and H. V. Dijk, chap. 6, pp. 221–292. Oxford University Press.

LIU, J. S. (1994): "The Collapsed Gibbs Sampler in Bayesian Computations with Applications to a Gene Regulation Problem," *Journal of the American Statistical Association*, 89(427), 9580–9660.

LIU, J. S., W. H. WONG, AND A. KONG (1994): "Covariance Structure of the Gibbs Sampler with Applications to the Comparisons of Estimators and Augmentation Schemes," *Biometrika*, 81(1), 27–40.

MASSY, W. F. (1966): "Order and Homogeneity of Family Specific Brand-Switching Processes," *Journal of Marketing Research*, 3(1), 48–54.

MCLACHLAN, G., AND D. PEEL (2000): *Finite Mixture Models*, Wiley Series in Probability and Statistics. John Wiley & Sons.

MOREIRA, M. J. (2003): "A Conditional Likelihood Ratio Test for Structural Models," *Econometrica*, 71, 1027–1048.

NEAL, R. M. (2000): "Markov Chain Sampling Methods for Dirichlet Process Mixture Models," *Journal of Computational and Graphical Statistics*, 9(2), 249–265.

NEWTON, M. A., AND A. E. RAFTERY (1994): "Approximate Bayesian Inference with the Weighted Likelihood Bootstrap," *Journal of the Royal Statistical Society, Series B*, 56(1), 3–48.

NORETS, A., AND J. PELENIS (2011): "Bayesian Modeling of Joint and Conditional Distributions," discussion paper, Princeton University.

OSBORNE, M. (2007): "Consumer Learning, Switching Costs, and Heterogeneity: A Structural Examination," discussion paper, Economic Analysis Group, Department of Justice.

QUANDT, R. E., AND J. B. RAMSEY (1978): "Estimating Mixtures of Normal Distributions and Switching Regressions," *Journal of the American Statistical Association*, 73(364), 730–738.

R Development Core Team (2012): *R: A Language and Environment for Statistical Computing*. R Foundation for Statistical Computing, Vienna, Austria.

ROSSI, P. E. (2012): *bayesm: Bayesian Inference for Marketing/Micro-Econometrics*, 2.2-5 ed.

ROSSI, P. E., AND G. M. ALLENBY (2011): "Bayesian Applications in Marketing," in *The Oxford Handbook of Bayesian Econometrics*, ed. by J. Geweke, G. Koop, and H. V. Dijk, chap. 8, pp. 390–438. Oxford University Press.

ROSSI, P. E., G. M. ALLENBY, AND R. E. MCCULLOCH (2005): *Bayesian Statistics and Marketing*. John Wiley & Sons.

ROSSI, P. E., R. E. MCCULLOCH, AND G. M. ALLENBY (1996): "The Value of Purchase History Data in Target Marketing," *Marketing Science*, 15(4), 321–340.

SEETHARAMAN, P. B. (2004): "Modeling Multiple Sources of State Dependence in Random Utility Models: A Distributed Lag Approach," *Marketing Science*, 23(2), 263–271.

SEETHARAMAN, P. B., A. AINSLIE, AND P. K. CHINTAGUNTA (1999): "Investigating Household State Dependence Effects across Categories," *Marketing Science*, 36(4), 488–500.

SETHURAMAN, J. (1994): "A Constructive Definition of Dirichlet Priors," *Statistica Sinica*, 4, 639–650.

SHUM, M. (2004): "Does Advertising Overcome Brand Loyalty? Evidence from the Breakfast-Cereals Market," *Journal of Economics and Management Strategy*, 13(2), 241–272.

STEPHENS, M. (2000): "Dealing with Label-Switching in Mixture Models," *Journal of the Royal Statistical Society, Series B*, 62, 795–809.

TADDY, M. A., AND A. KOTTAS (2010): "A Bayesian Nonparametric Approach to Inference for Quantile Regression," *Journal of Business and Economic Statistics*, 28(3), 357–369.

VILLANI, M., R. KOHN, AND P. GIORDANI (2009): "Regression Density Estimation Using Smooth Adaptive Gaussian Mixtures," *Journal of Econometrics*, 153(2), 155–173.

Index

Clustering, 46–49

Dirichlet distribution, 17
Dirichlet process; base measure, 60; construction, 60–64; limit of finite mixtures, 67; marginal distribution, 62; number unique values, 72; Polya Urn representation, 63; stick-breaking representation, 68; tightness parameter, 60
Dirichlet-Multinomial conjugacy, 61

E-M algorithm, 10–13
Examples; Acemoglu IV, 145; bivariate non-parametric regression, 101; Card IV, 146; Chi-squared distribution, 39, 81; Comparison of DP and finite mixtures, 88; Gelman-Meng distribution, 40; Label-Switching, 30; log-normal distribution, 43, 86; outliers, 57; random coefficient logit model, 161–186; random coefficient logit model data, 168; univariate non-parametric regression, 97

Finite mixture of normals, 2

Gibbs sampler; collapsed, 49–58; Dirichlet process, 70; IV model with DP errors, 130; marginalized, 49–58; mixture of DP, 78; normal IV model, 125; unconstrained, 25–29

heterogeneity and state dependence, 172
Homescan data, 109

Inverted Wishart distribution, 19
IV model; normal case, 124

Kernel smoothing, 22, 39, 41

Label-Switching, 29
Likelihood function; finite mixture of normals, 6

Non-parametric expenditure function, 108
Non-parametric regression, 90–114; conditional mean, 94; conditional quantiles, 94; discrete dependent variable, 104–107

Posterior model probabilities, 166
Predictive density; multivariate normal model, 53
Predictive distribution of data, 79
Prior assessment; DP base measure hyper-parameters, 77; DP parameters, 72–78; DP tightness parameter, 74; finite mixture of normals, 16–25; IV model, 129
Prior sensitivity; DP base measure hyper-parameters, 81; DP tightness parameter, 83, 84; finite mixture, 38; semi-parametric IV model, 140

Random coefficient models, 152–186

Saddle point, 8
Scale mixture, 1

Scaling data, 21, 80
Semi-parametric methods, 115–151;
   binary choice model, 192; binary
   choice with IV, 193; IV estimation
   performance, 138; IV interval
   coverage, 133; IV model, 122–151;
   IV model sampling experiments,
   131–145; IV sampling experiment

design, 131; random coefficient
   logit models, 157; sample selection
   model, 193
Semi-parametric regression,
   115–121
State dependence, 163; and
   autocorrelation, 178
Sufficient statistics; updating, 53